Green Energy and Technology

More information about this series at http://www.springer.com/series/8059

Lorenzo Battisti · Mosè Ricci
Editors

Wind Energy Exploitation in Urban Environment

TUrbWind 2017 Colloquium

Springer

Editors
Lorenzo Battisti
University of Trento
Trento
Italy

Mosè Ricci
University of Trento
Trento
Italy

ISSN 1865-3529 ISSN 1865-3537 (electronic)
Green Energy and Technology
ISBN 978-3-030-09111-8 ISBN 978-3-319-74944-0 (eBook)
https://doi.org/10.1007/978-3-319-74944-0

Softcover re-print of the Hardcover 1st edition 2018

Printed on acid-free paper

This Springer imprint is published by Springer Nature
The registered company is Springer International Publishing AG
The registered company address is: Gewerbestrasse 11, 6330 Cham, Switzerland

Preface

The volume contains the written versions of most of the contributions presented during the *TUrbWind Colloquium—Research and innovation on wind energy exploitation in urban environment* organized by the University of Trento (IT) that took place in Riva del Garda, Italy, from June 15 to 16, 2017.

The purpose of this colloquium was to discuss the state-of-the-art and the recent developments on wind energy exploitation in urban and suburban area. The urban environment offers a peculiar wind resource, highly variable and extremely turbulent. In order to harvest wind energy in an efficient and durable manner, urban wind turbines have to be properly designed. At the same time, this environment is where the human energy demand is mostly concentrated. Integration of wind turbines on existing or dedicated buildings or, where possible, as free-standing installation in public areas, requires a different design approach compared to open areas. Such issues drive opportunities for innovation and development.

The TUrbWind Colloquium was intended to provide a forum for experts on wind energy exploitation in urban areas to present and discuss the state of the art of the technology, exchange ideas and get informed about the latest research trends in the domain. In particular, this colloquium aimed to bring the research community together for the first time under a dedicated event on the urban wind turbines and their integration in the anthropic areas.

We would like to thank all participants for their contributions to the colloquium program and for their contributions to these proceedings. We also acknowledge the important contributions of Prof. Gerard van Bussel and Dr. Josè Alfredo Ramirez in opening the colloquium program. Our special thanks go to colleagues Dr. Ignacio Cruz and Dr. David Wood, for their assistance during the colloquium, and to the members of the Scientific Board. It is our pleasant duty to acknowledge the support of the InVento promoters in the overall organization of the colloquium.

Trento, Italy
November 2017

Lorenzo Battisti
Mosè Ricci

Contents

Experimental and Numerical Wind-Resource Assessment of an University Campus Site

A. El Bahlouli and J. Bange

Abstract During a 3 year research project funded by the local government of Baden-Württemberg, Germany, the potential of wind-energy production was studied at the university campus of Tübingen, a town in the south-west of Germany. The 3D wind field was studied both experimentally and numerically in order to identify optimal locations for small wind turbine installation. Within the scope of this project, a full-scale field experiment and RANS (Reynolds Averaged Navier-Stokes) models were applied in order to yield a better understanding of the airflow around the buildings. We validate our CFD predictions of the flow field with wind-speed measurements using ultrasonic anemometers at several stations within the campus. The simulation results (in direct comparison with the measured data) improved greatly when trees were explicitly considered using a simple canopy model at the inflow boundary. This study is intended to support and guide the next steps of the wind resources assessment at similar sites. We gladly offer our site, instrumentation and (simulated and measured) data to other groups that perform urban wind energy studies.

Keywords Wind flow · Urban environment · RANS · Anemometers

1 Introduction

Several international and national policies and grant programmes are encouraging the use of renewable sources of energy, among which are small-scale wind-energy technologies. Small wind turbines (SWT, less than 100 kW as defined by the World Wind Energy Association WWEA) have shown a remarkable growth in use during recent years mainly in China and the USA. In 2015, an increase of 14% compared with the previous year has been registered for the small wind capacity installed worldwide. This market is expected to reach a steady growth rate of 20% from 2015 to 2020 as reported by the WWEA in their 'Small Wind World Report Summary'

A. El Bahlouli · J. Bange (✉)
Environmental Physics, University of Tübingen, Hölderlinstr. 12,
72074 Tübingen, Germany
e-mail: jens.bange@uni-tuebingen.de

L. Battisti and M. Ricci (eds.), *Wind Energy Exploitation in Urban Environment*,
Green Energy and Technology, https://doi.org/10.1007/978-3-319-74944-0_1

(www.wwindea.org). These small wind energy technologies were in the focus of a 3 year research project at the University of Tübingen, in the south-west of Germany, as a pioneer study in sustainability and environmental management in the region. The university and the university hospitals of Tübingen are using around 20% of their electricity from renewable sources, at the time of writing. The aim is to increase this share to 35% by 2020. A large and efficient photo-voltaic system is already in service but no wind energy resources are there to complement them. Thus first aim of the study was a detailed wind resource site assessment before locations could be identified where wind-energy resources are possible.

Historically, wind energy technologies were built in open areas with favourable wind conditions, the basic condition in order to harvest significant wind power. The atmospheric boundary layer (ABL) behaves differently in urban environments compared to the flow in open areas. The surface geometry in urban areas is significantly more complex than for open spaces and has a critical influence on wind flow at the micro-meteorological scale. Several studies have shown that urban areas have relatively low average wind speeds [8, 18], but turbulence can be much stronger due to the larger shear and strong local convection in the presence of obstacles and heated concrete, causing extreme wind events of relatively short duration (strong gusts). The presence of buildings disturbs the wind flow by generating zones of wind acceleration, channelling, blocking, re-circulation and increasing turbulence. Wind turbines installed in urban areas should take the features of the urban ABL into account. Such wind turbines will experience frequent wind changes, a large amount of turbulence, gusts and a significant vertical wind component (in conflict with the usual design of wind turbines). All this reduces the expected life time of the turbines, and can increase the risk of breakdowns due to the large amount of fatigue load on the structure. It is therefore necessary to perform more measurements and a detailed wind resource assessment in the potential site to optimise the placement of SWT.

In situ measurements are the only way to characterise the wind flow at the potential sites, adequately. Such measurements are a good way to quantify the wind resource as it accounts for all environmental parameters and weather fluctuations. But these measurements are often limited to a meteorological station installed at a single location that is usually not be fully representative of the flow over the site. And such measurements have to be performed over a long period to get reliable statistics [2]. All these constraints make in situ measurement in the urban environment relatively limited. Thus many studies are conducted in wind tunnel experiments to measure wind flows around building models in a stationary flow.

During the past two decades, many CFD (Computational Fluid Dynamics) methods have been applied in wind project studies. CFD tools offer considerable advantages and provide detailed information of the flow in the whole calculation domain. The use of numerical models for wind engineering application known as Computational Wind Engineering (CWE) increased significantly during the last decade [3]. It is widely recognised that the availability of many physical and numerical parameters in these methods, which can be freely chosen by the user, can lead to wrong assumptions or numerical errors. Typically, a user has to choose the approximation form of the governing equations, the turbulence models, the discretisation schemes, the computational domain, etc. To overcome this problem, the wind-energy community published several best practice guideline documents and recommendations on the use of CWE, as in [4, 6, 17].

Since the power output from a wind turbine is proportional to the wind speed cubed, accurate estimation of the campus wind speed is essential for accurate power output estimates. In order to study the wind-energy resources at the university site, a detailed wind resource site assessment was planned. Here, we like to present the first step of the wind resources assessment study which include a CFD validation with high-resolution wind data measured on-site.

After 3 years, the project stopped since the funding expired. But we gladly offer our site, instrumentation and (simulated and measured) data to other groups that perform urban wind energy studies.

2 Methodology

2.1 The Campus Site Specification

The project focuses on one area, called Morgenstelle, where the average wind speed is estimated to be the highest among the university buildings. The chosen investigation site is located in outlying districts of the city, on top of a hill. The site has a mix of shallow and high buildings with three of them approximately 50 m height. Previous simulation using the meso-scale model METRAS PC [13] on a 500 m $\times$ 500 m grid spacing, indicated the prevailing wind direction at the side is south-west, as shown in Fig. 1. The majority of the buildings are oriented in a south-west direction, as well. Upstream the site, i.e. the western part of the campus, is covered with a forest. This area lies in a valley that influences the wind flow at Morgenstelle directly, in case of west or south-west wind direction.

2.2 The Full Scale Experiment

A common technique to calculate the energy potential includes wind measurements at least for one annual cycle. Thus, six stations consisting of three-dimensional ultrasonic anemometers (RM Young 81000) have been installed at the site since April, 2014. In order to measure the distortion of the flow due to building effects, different kind of locations for the instruments were chosen, such as roofs, balconies or on the ground (see Fig. 2).

The anemometers are mounted on 3 m masts (the maximum height authorised), except for the balcony station which has a 1 m horizontal boom. Around the anemometers, complementary equipment (e.g. power supply and data acquisition systems) were installed. Due to different conditions at the chosen measurement

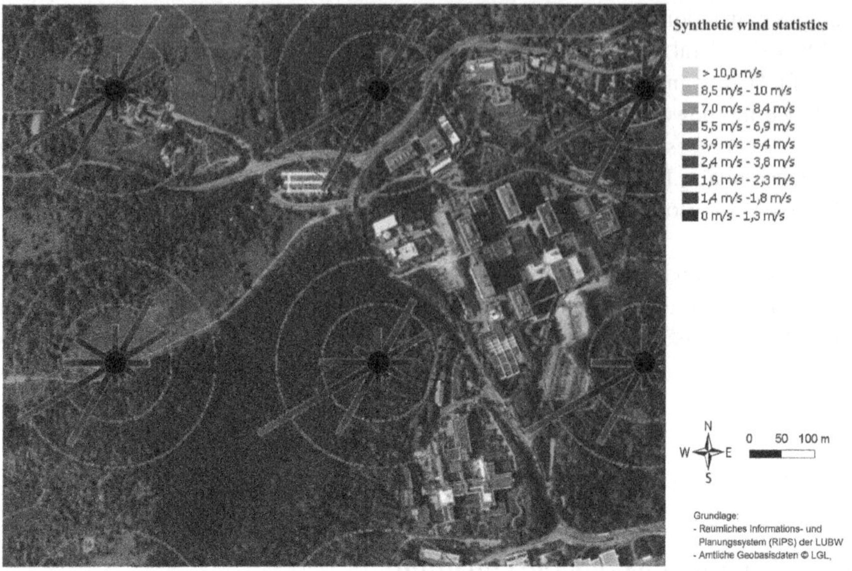

Fig. 1 Aerial view of the site and wind statistics from the METRAS-PC model. *Source* www.udo.lubw.baden-wuerttemberg.de

locations, at some stations data loggers (Campbell Scientific CR800) operating on an external battery are used. For the remaining stations, grid powered systems driven by a microcomputer (Raspberry Pi) were developed and built. The anemometers provide the three wind components and the 'sonic' (close to virtual) temperature with high temporal resolution (10 Hz). Therefore, the stations generate a large amount of data, around 300 MByte per day. Thus, a database management using a Structured Query Language (SQL) became necessary.

2.3 The Simulation Set-Up

In the current study, the calculations are carried out only for the prevailing wind direction of 240°. The simulations are performed using the open source C++ code OpenFOAM (openfoam.org). Here, a Reynolds Averaged Navier-Stokes (RANS) approach is used to investigate the flow around and through the site. The computational geometry and domain are built in accordance with the best practice guidelines by [6, 17]. Considering these guidelines, the resulting computational domain has dimensions of $L \times W \times H = 750\,\text{m} \times 1250\,\text{m} \times 350\,\text{m}$. The size of the entire computational domain in the vertical, lateral and flow directions is not limited to the region of interest but includes the surroundings (Fig. 3).

Fig. 2 Photographs of measurement station. 1: balcony, 2: garden, 3: canteen, 4: 'A' building, 5: lecture hall and 6: heating plant

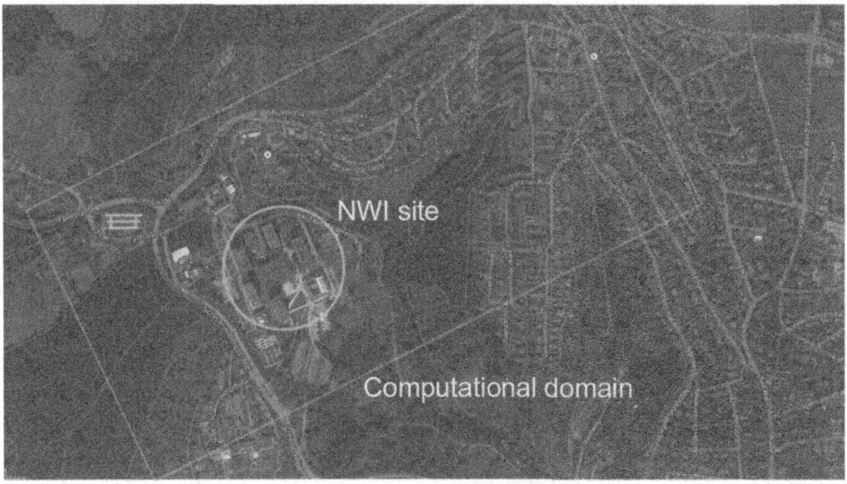

Fig. 3 The computational domain around the Morgenstelle (NWI site)

In the model, the ground has been simplified by assuming it to be flat. The computational grid has been created with the SnappyHexMesh utility (the OpenFOAM mesh generator) and contains approximately 8 millions cells with a resolution of 1.5 m near the area of interest. The simulations were carried out with the k-ε model of [9] and the simpleFoam solver, which is a steady state, incompressible solver. The inlet uses Dirichlet conditions for the velocity U, the turbulent kinetic energy k and the dissipation rate ε with a log-law velocity profile using the OpenFOAM libraries as in [14] with a reference wind speed of 5 m/s at 10 m height. Flow is considered to be fully developed in the outlet, thus we apply a Neumann zero gradient condition for all variables, except for pressure. For this last variable a Neumann zero gradient condition is assumed. For the ground, a no-slip condition is set for the wind field and zero gradient condition is used for the pressure. The wall shear stress is computed by the standard OpenFOAM wall functions with a homogeneous roughness height of 0.03 m. A slip condition is used for the sides (parallel to the flow direction) and the top of the domain.

2.4 The Canopy Model

Figure 1 shows that the site is located at the edge of a forest at the in-flow boundary for westerly winds. Thus, it was necessary to include the effect of the forest on the wind flow in our study. In many studies in urban area, details like trees are neglected or implicitly considered in the roughness parameter. E.g. [12, 15] used an explicit representation with a canopy model for reproducing the aerodynamic effects of the trees in urban area. These studies showed that using the implicit approach leads to small effects of trees in most of the urban domain simulated. Thus we decided to consider the effect of the forest at the in-flow boundary using an explicit approach. To include the impact of the forest on the spatial distribution of wind speed, we use a canopy model based on the work of [10] where the vegetation is represented by a leaf area density (LAD) profile. To consider the effects of the canopy of the flow, the incompressibly filtered Navier-Stokes equations have been modified following the ideas of [10]. This method adds an additional drag force D_i in the x_i direction generated by the forest in the following way:

$$D_i = -C_d \, \mathrm{LAD}(z) \; V \; U_i \tag{1}$$

where C_d is a constant drag coefficient, V is the mean wind speed, U_i the local wind velocity in i direction, and LAD the leaf area density at height z. The drag force D_i is added in the momentum equation but also in the kinetic-energy equation and the dissipation equation. The drag coefficient C_d is a parameter which links the canopy architecture with its aerodynamic behaviour. In two mixed forests of the Italian Alps, [5] investigated the drag coefficient experimentally and found mean values around 0.09 ± 0.06 and 0.12 ± 0.06. Having a mixed forest in our site, we decided to continue our study with a value of 0.15.

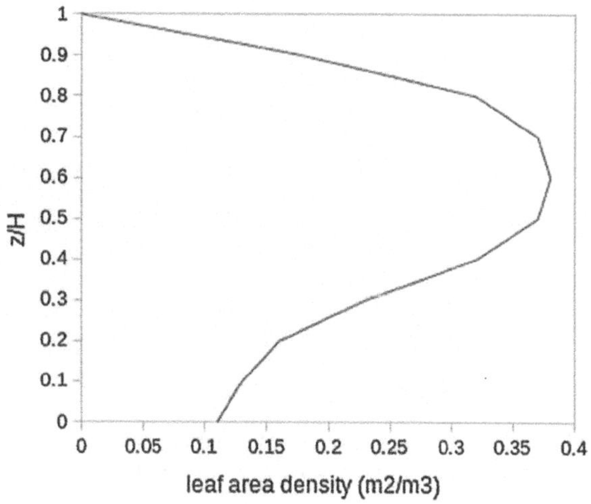

Fig. 4 Vertical distribution of leaf area density

The LAD (m^2 leaf area per m^3 canopy volume) is only a function of height in this model and depends on species, developmental stage, leaf season, etc. This function varies over height, since the leafs are not equally distributed along the trunk height. The LAD is derived from the leaf area index LAI as follows:

$$\mathrm{LAI} = \int_z^H \mathrm{LAD}\, dz \tag{2}$$

A LAI of 5 is chosen after a literature survey, e.g. [7, 11, 16]. Figure 4 shows the resulting leaf area density profile as a function of z/H, where H is the tree height of about 22 m, in our study.

3 Results

3.1 Verification of the Numerical Simulation

Our approach to compare measurements with simulations is somewhat upside-down. We first conducted a simulation and then identified periods in our database that agree with the boundary conditions of the simulation. In order to compare the RANS simulations with our measurement stations, we chose the station mounted on the highest building near the inflow, named 'A-building', also shown in Fig. 5, as a first reference. In case of south-west wind, this station is not in the wind shadow of any other building. Also the instruments are located at the leading edge in the centre of the roof. Thus, the wind direction at this station is assumed to be identical with the inlet profile, which equals 240° due to our simulation settings. Knowing the other

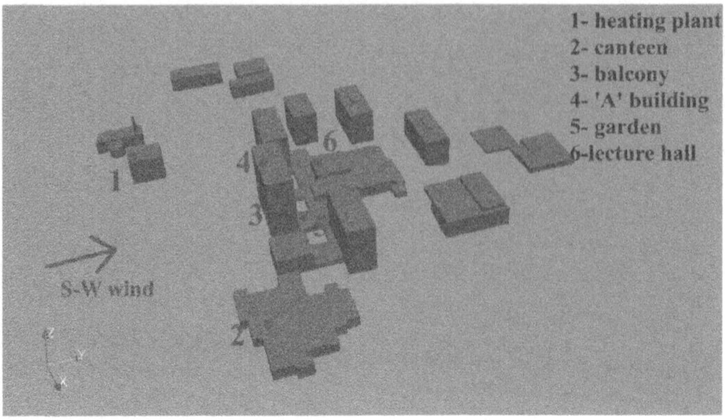

Fig. 5 Top: 3D perspective of Morgenstelle site, buildings are shown in grey, and idealised smoothed topography is shown in brown, the station network is shown in red (color figure online)

station's location, we can easily find the corresponding coordinates in our simulation and extracted the simulated values of the mean wind speed to be expected at the remaining stations.

The simulation gives a mean velocity of 5.40 m s^{-1} at the 'A-building' station. In the SQL database, we look for 30 min periods fulfilling both following conditions at the A-building station:

1. The mean wind speed is 5.40 ± 0.1 m s^{-1}.
2. The wind direction is $240 \pm 10°$.

Finally, 28 data sets were chosen in order to have various weather conditions (day and night time, cold or warm days), and listed in Table 1.

The scatter plot in Fig. 6 shows the simulated velocities without the application of our canopy model (Sect. 2.4), matched by the 30 min-average wind speed from the stations. The spread in the experimental data (vertical axis) is in the order of 2 m s^{-1} and almost constant for all the stations. The plot shows that the experimental data don't match the simulation well. The stations 5 (garden) and 3 (balcony) are the ones closer to the line of best fit. This can be explained by the fact that station 5 and station 3 are located behind or in a corner of a building (Fig. 5). Thus the data shows mainly the effect of the building regardless of the incoming wind field. Stations 1 (heating plant) and 2 (canteen) are the ones directly influenced by the inlet profile. Here, measurements show much smaller wind speeds compared to the simulation. In general, the simulation over-estimated the real wind speeds significantly, probably due to the insufficient modelling of the incoming wind profile.

Our first simple approach of a flat terrain with a homogeneous roughness height is not suited to the problem. Mainly because the site is surrounded by a significant number of trees in the upstream area. We need to adjust the upstream area of the campus by adding the canopy model (Sect. 2.4) as we expect that the forest has an impact on the spatial distribution of wind speed.

Table 1 Data sets used for comparison. Units for the wind speed are m s^{-1}

Date time	A-Building	Canteen	Heating plant	Balcony	Garden	Lecture hall
03/05/15 12:00	5.43	2.09	1.81	4.29	2.10	–
03/05/15 20:30	5.42	1.50	1.22	3.75	1.51	–
06/05/15 20:30	5.42	1.99	2.23	4.57	1.43	1.87
06/05/15 16:00	5.42	2.13	1.67	3.31	1.83	2.73
06/07/15 01:30	5.43	1.48	1.77	3.76	1.27	2.09
06/07/15 03:30	5.39	1.47	1.80	3.83	1.03	2.01
07/07/15 20:00	5.41	1.01	1.61	3.39	0.82	1.61
07/07/15 04:30	5.34	0.70	1.22	3.13	1.04	1.75
08/07/15 03:30	5.43	1.66	2.24	1.12	4.09	2.02
08/07/15 12:30	5.43	2.36	2.06	1.45	4.38	2.43
15/11/15 20:00	5.44	2.04	1.25	3.57	2.21	2.08
15/11/15 22:00	5.42	1.52	0.89	3.55	2.02	0.97
17/11/15 02:30	5.43	1.13	1.43	3.31	1.09	1.24
17/11/15 04:00	5.45	1.84	1.85	4.29	1.77	1.12
25/12/15 10:00	5.41	1.59	1.32	2.91	1.01	1.88
25/12/15 20:30	5.41	1.37	1.19	3.72	1.03	0.98
29/12/15 05:00	5.36	1.78	1.2	1.46	1.31	1.93
29/12/15 09:00	5.33	2.11	1.11	3.34	0.76	3.34
04/01/16 06:00	5.44	1.47	1.43	3.81	1.28	1.23
04/01/16 18:30	5.42	1.32	1.18	3.09	1.75	1.80

(continued)

Table 1 (continued)

Date time	A-Building	Canteen	Heating plant	Balcony	Garden	Lecture hall
05/01/16 14:30	5.44	1.53	1.29	3.12	1.28	1.94
05/01/16 19:30	5.42	1.42	1.23	3.36	1.30	1.14
05/02/16 09:00	5.44	1.54	1.44	3.70	1.58	1.85
05/02/16 15:30	5.48	1.75	1.77	3.97	1.81	2.20
06/02/16 08:00	5.40	2.41	1.41	4.74	1.11	1.96
06/02/16 14:00	5.44	1.56	1.28	2.92	1.36	1.47
10/02/16 03:00	5.45	1.75	1.69	4.13	1.68	2.24
10/02/16 05:30	5.45	1.95	1.85	4.07	1.83	1.92

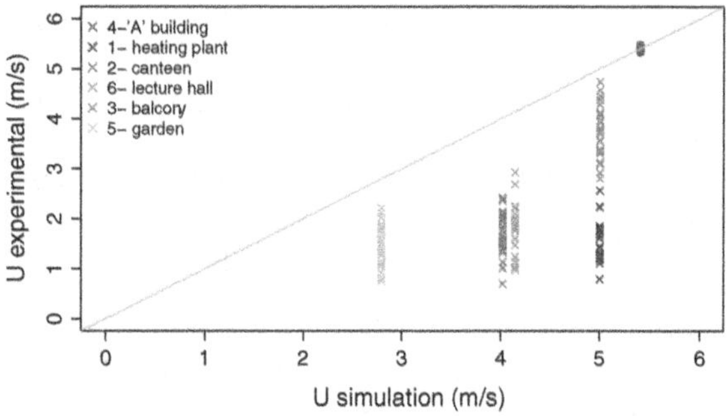

Fig. 6 Scatter plot of measured wind speed against simulated wind speed

Results of the simulation using the canopy model are shown in the Fig. 7. Now, all the simulated wind velocities are decreased compared to the previous case, but meet the experimental data much better. However, station 1 (heating plant) still shows a significant difference between the experimental data and simulation result. It should be noted that this station is located only 30 m away from the forest and thus is very sensitive to the forest representation in the model, which is—so far in this study and in contrast to reality—horizontally homogeneous. An even more detailed forest model might give further improvements. However, the results illustrate the importance of having a realistic inlet profile by considering, for example, the forest canopy.

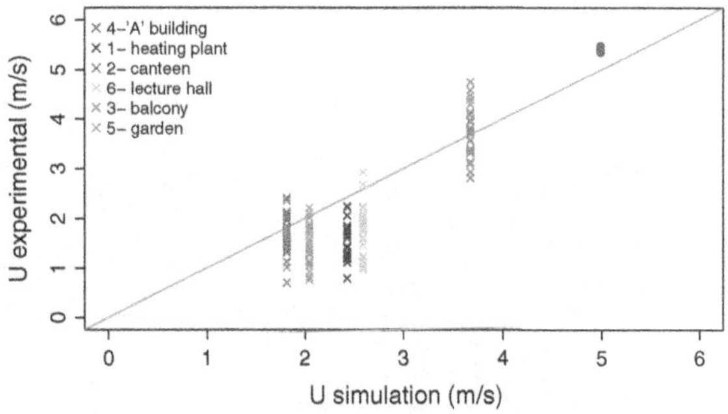

Fig. 7 Scatter plot of measured wind speed against simulated wind direction with a canopy forest

3.2 *Accelerated Flow Around and Above the Buildings*

The following 3D figures where obtained using the post-processing tool Paraview [1] which helps to identify the areas where the velocity is increased (in comparison to the inlet velocity) and to visualise the air flow, both in three dimensions. Figure 8 shows the velocity streamlines. Air flow patterns around buildings is quite complex. At the windward surface of the buildings, where the wind first impacts the buildings, the wind stream splits above and around the sides of the buildings. The air flow accelerates around the corners of the building causing a high speed region of the corner and a weak wind region on the sides of the buildings.

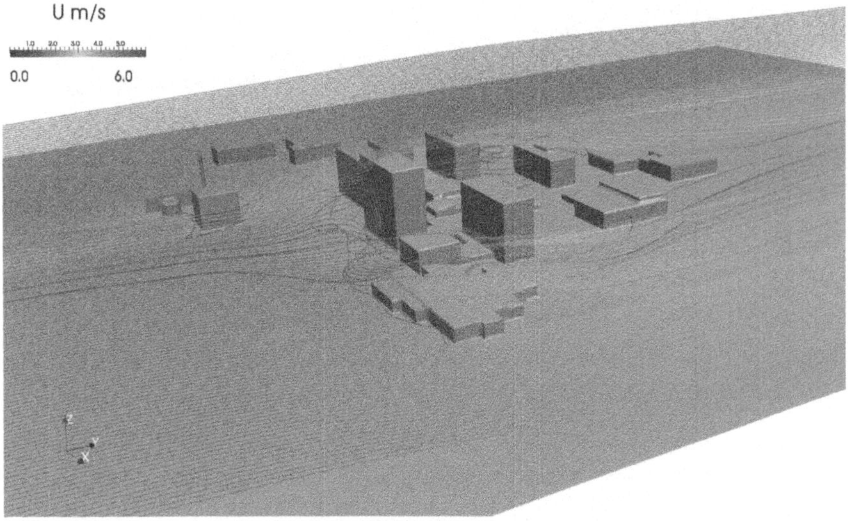

Fig. 8 Velocity stream lines at Morgenstelle campus. Colours indicates wind speed close to zero (blue) up to 6 m/s (red) (color figure online)

Fig. 9 Velocity contour-plots at height of: (top) 2 m, (centre) 20 m and (bottom) 50 m

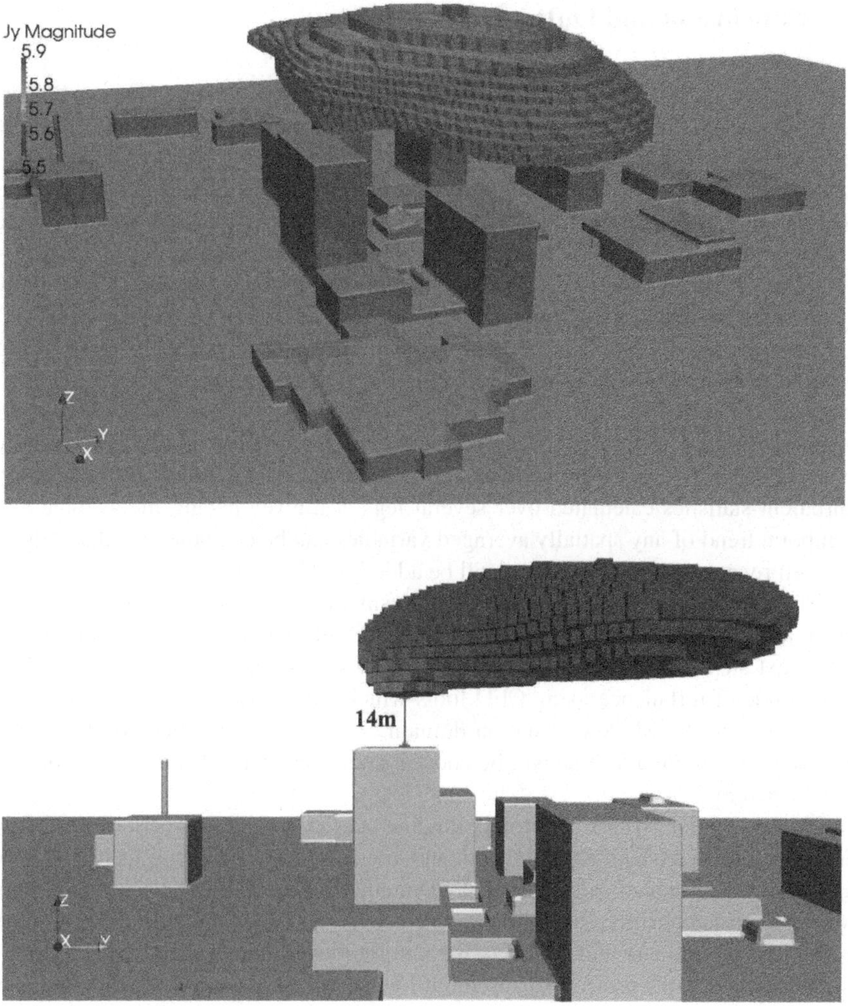

Fig. 10 Model cells with a minimum velocity increase of 10%. Top: 3D view, and bottom: side view

The velocity contour-plots at different heights (Fig. 9) illustrate this phenomena. The wind speed is also increasing with altitude as expected (due to the logarithmic wind profile at the inlet).

The blue area in Fig. 10 (top) represents the cells with a minimum horizontal wind speed increase of 10%, which corresponds to a minimum of 5.5 m/s. This area is located over the 'A'-building and—due to its increased wind speed—is attractive for wind-energy production. To reach this area, a mast of minimum 14 m height with a turbine on top would be necessary, as seen in Fig. 10 (bottom).

4 Conclusion and Outlook

This paper reports the first step of the feasibility of wind power utilisation in a university campus and find out how to model the wind flow in the built environment more effectively. Two RANS simulations were performed for the flow over and through the university buildings. Including a canopy model improved the simulation results significantly, demonstrating the importance of having a realistic inlet profile. While these simulations represent first results, they illustrate the potential for future investigations of canopy turbulence. And a more realistic ground model (and horizontally heterogeneous inlet) would improve the results even more.

We planned to measure the incoming wind flow over the forest in more detail using a small unmanned aerial vehicle UAV [19, 20]. The research UAV MASC (Multi-purpose Airborne Sensor Carrier) can resolve turbulence fluctuation of wind and temperature up to 30 Hz. Legs (straight and level flight sections) of approximately 1 km total length can be flown over the forest at various altitudes. By comparing turbulent statistics calculated over several legs at a given height, the average and temporal trend of any spatially averaged variables can be computed for that height. Therefore, new experimental data will be added for CFD validation.

The first step of this wind resource assessment helped us to identify important parameters in our model which were validated with our long term measurements. The next step will consist of identifying spots with higher wind speeds and low magnitude of turbulence using CFD tools which can provide detailed information of the flow in the whole calculation domain. After picking the best locations, the optimal type of wind turbine will be chosen, and an estimation of the power output can be given.

However, after 3 years the project funding ceased. During this period, six turbulence stations were built and mounted, and a measurement data base including user interface was designed. Also various CFD methods were tested and simulation runs were performed on different computation clusters. It is understood that the presented results are preliminary and that there is room for plenty of improvements and extensions. Currently, the project is paused until we applied successfully for a new grant or/and we found a new partner who wants to use the installations and the database at the urban wind field research site Morgenstelle.

Acknowledgements The authors thank the Ministry of Science, Research and the Arts of the State of Baden-Württemberg for funding this study, the Geographical Institute of the University of Tübingen for providing terrain data, and the Technical Building Management (TBA) of the University of Tübingen for assisting the experimental part. The computational resources were provided by the bwGRiD Cluster at the University of Tübingen.

References

1. Ayachit, U.: The paraview guide: a parallel visualization application (2015)
2. Bailey, B.H., McDonald, S.L., Bernadett, D., Markus, M., Elsholz, K.: Wind resource assessment handbook: fundamentals for conducting a successful monitoring program. Tech. rep., National Renewable Energy Lab., Golden, CO (US); AWS Scientific, Inc., Albany, NY (US) (1997)
3. Blocken, B.: 50 years of computational wind engineering: past, present and future. J. Wind Energy Ind. Aerodyn. **129**, 69–102 (2014)
4. Blocken, B., Gualtieri, C.: Ten iterative steps for mode development and evaluation applied to computational fluid dynamics for environmental fluid mechanics. Env. Model. Softw. **33**, 1–22 (2012)
5. Cescatti, A., Marcolla, B.: Drag coefficient and turbulence intensity in conifer canopies. Agric. For. Meteorol. **121**, 197–206 (2004)
6. Franke, J., Hellsten, A., Schlnzen, H., Carissimo, B.: The COST 732 best practice guideline for CFD simulation of flows in the urban environment: a summary. Int. J. Env. Pollut. **44**, 419–427 (2011)
7. Greens, S., Grace, J., Hutchings, N.: Observations of turbulent air flow in three stands of widely spaced sitka spruce. Agric. For. Meteorol. **74**, 205–225 (1996)
8. Landberg, L., Myllerup, L., Rathmann, O., Petersen, L., Hoffmann Jrgensen, B., Badger, J., Gylling Mortensen, N.: Wind resource estimation—an overview. Wind Energy **6**, 261–271 (2003)
9. Launder, B.E., Spalding, D.: The numerical computation of turbulent flows. Comput. Meth. Appl. Mech. Eng. **3**, 269–289 (1974)
10. Liu, J., Chen, J., Novak, M.: k-epsilon modelling of turbulent air flow downwind of a model forest edge. Bound.-Layer Meteorol. **77**, 21–44 (1996)
11. Meier, I., Leuschner, C.: Leaf size and leaf area index in fagus sylvatica forests: competing effects of precipitation, temperature, and nitrogen availability. Ecosystems **11**, 655–669 (2008)
12. Mochida, A., Tabata, Y., Iwata, T., Yoshino, H.: Examining tree canopy models for CFD predicyion of wind environment at pedestrian level. J. Wind Energy Ind. Aerodyn. **96**, 1667–1677 (2008)
13. Rau, I.M., Bigalke, K.: Synthetische Windstatistiken Baden-Württemberg Hinweise für Anwender. Landesanstalt für Umwelt, Messungen und Naturschutz Baden-Württemberg (2007)
14. Richards, P., Hoxey, R.: Appropriate boundary conditions for computational wind engineering models using the k-epsilon turbulence models. J. Wind Energy Ind. Aerodyn. **46–47**, 145–153 (1993)
15. Salim, M.H., Schlnzen, H.K., Grawe, D.: Including trees in the numerical simulations of the wind flow in urban areas: should we care? J. Wind Energy Ind. Aerodyn. **144**, 84–95 (2015)
16. Thimonier, A., Sedivy, I., Schleppi, P.: Estimating leaf area index in different types of mature forest stands in switzerland: a comparison of methods. Eur. J. For. Res. **129**, 543–562 (2010)
17. Tominaga, Y., Mochida, A., Yoshie, R., Kataoka, H., Nozu, T., Yoshikawa, M., Shirasawa, T.: AIJ guidelines for practical applications of CFD to pedestrian wind environment around buildings. J. Wind Energy Ind. Aerodyn. **96**, 1749–1761 (2008)
18. Walker, S.: Building mounted wind turbines and their suitability for the urban scalea review of methods of estimating urban wind resource. J. Energy Build. **43**, 1852–1862 (2013)
19. Wildmann, N., Hofsäß, M., Weimer, F., Joos, A., Bange, J.: MASC—a small remotely piloted aircraft (RPA) for wind energy research. Adv. Sci. Res. **11**, 55–61 (2014)
20. Wildmann, N., Rau, G., Bange, J.: Observations of the early morning boundary-layer transition with small remotely-piloted aircraft. Bound.-Layer Meteorol. **157**(3), 345–373 (2015)

Measurements of Semi-urban Gust Factors for Wind Load Determination

O. L. C. Antonio and D. H. Wood

Abstract Wind turbine and building codes used to design photovoltaic installations among other structures, typically use a gust factor to convert wind speeds averaged over 10 min (IEC standards for wind turbines) or 1 h (many building codes) into 3 s averages for determining the extreme wind loads. The IEC standard for small wind turbines assumes the gust factor has a universal value of 1.4, which is close to that used in many international wind loading standards. Ultrasonic wind speed measurements at a height of 50 m in a semi-urban location in north Calgary were sampled at 100 Hz and analyzed to determine the gust factor in a novel manner. 692 h of measurements were obtained and fitted to the generalized extreme value probability distribution function. We find gust factors as low as 1.26 at the 99% confidence interval, corresponding to 3 standard deviations from the mean value. Since the square of the gust factor is used in estimating extreme loads, this represents a 19% reduction in these loads compared to the values mandated by the IEC standard.

Keywords Gust factor · Urban wind flow · Wind energy · Anemometry
Wind loads

1 Introduction

Structural safety and wind turbine standards like IEC 61400-2 [1] use a gust factor (GF) in determining extreme wind loads. For example, Load Case H in IEC 61400-2 requires a small turbine to be safe for a "3 s gust wind speed" of V_{e50} which is 1.4 times V_{ref}, the reference wind speed defined as a 10 min average. In this case $GF = 1.4$. Most wind speed data is averaged over either 10 min or 1 h and the use of a GF can be viewed as a "correction" to obtain averages over the minimum time period sufficient to cause severe structural damage. By convention, this time is normally taken as 3 s, e.g. Holmes [2]. In determining wind loads, it is assumed, or

O. L. C. Antonio · D. H. Wood (✉)
University of Calgary, Calgary, AB T2N 1N4, Canada
e-mail: dhwood@ucalgary.ca

L. Battisti and M. Ricci (eds.), *Wind Energy Exploitation in Urban Environment*,
Green Energy and Technology, https://doi.org/10.1007/978-3-319-74944-0_2

must be assumed, that the *GF* is more general than the mean wind speed, but there is surprisingly little information available for it, and none that we are aware of from urban measurements. In this paper, *GF* is defined as the ratio of the 3 s to 10min average wind speed. However it can have many definitions, Antonio [3].

A Gaussian wind speed probability distribution function (PDF) is uniquely defined by its mean, $\bar{U}$, and standard deviation, σ_u. Then the *GF* can be related to the turbulence intensity $\sigma_u/\bar{U}$, by

$$GF = \left(\bar{U} + g\sigma_u\right)/\bar{U} = 1 + g\sigma_u/\bar{U} \tag{1}$$

where g is the peak factor, which in turn can be calculated as described in Davenport [4], Solari and Kareem [5] and Holmes [2]. Wang et al. [6] measured g and *GF* as a function of $\bar{U}$. Balderrama et al. [7] determined g for tropical cyclones on the south-east coast of the USA and obtained *GF* in the range of 1.46 and 1.50. *GF* was found to depend on σ_u, surface roughness, and terrain. Davis and Newstein [8] used regression to find *GF* as a function of $\bar{U}$, and found *GF* tended to a constant as $\bar{U}$ increased. Extreme wind speed estimation for wind energy applications is described by Anastasiades and McSharry [9] and Ishihara and Yamaguchi [10], who both use computer simulations to extend the wind speed database to estimate extreme values. These results suggest that the complexity of urban wind flow may require a range of values for *GF*, but as noted above, no serious study of urban *GF*s has been undertaken.

Equation (1) allows another determination of *GF*. The normal turbulence model in [1] specifies

$$\sigma_u/\bar{U} = 0.12 + 0.9/\bar{U} \tag{2}$$

Holmes [2], Table 3.4 gives $g = 2.5$ for a 3 s gust and 10 min averages. Thus

$$GF = 1.3 + 2.25/\bar{U} \tag{3}$$

These methods of determining *GF* are not the only possible ones, especially if *GF* can be assumed to be independent of $\bar{U}$, as this decouples the determination of *GF* from finding the extreme $\bar{U}$. Furthermore, it is well-known that wind speeds do not follow the Gaussian PDF and the more complex Weibull PDF is commonly used as a better approximation.

A specific aim of the present investigation was to determine whether *GF* is constant at high $\bar{U}$ for semi-urban settings as represented by our site. In addition, our mast is fitted with ultrasonic anemometers. These have become available only relatively recently and have very good frequency response, so that older methods of determining *GF*, based on measuring peak values, can be avoided. We decided to determine *GF* in a new way: through the Generalized Extreme Value (GEV) PDF, e.g. Kotz and Nadarajah [11]. The GEV is applicable for extreme value statistics rather than routine wind speed statistics. Nevertheless, it is a relatively simple PDF with three independent parameters that reduces to the widely used Weibull distribution for suitable parameter values.

The overall aim of the present paper is to use the GEV to analyze wind speed data to determine the *GF* and then to assess its dependence on $\bar{U}$. The next section introduces the GEV. The following section describes the measurement site and anemometer mast, and the data acquisition and processing, followed by a description of the application of GEV to the present data and the estimation of *GF*. The final section gives the conclusions.

2 The Generalized Extreme Value Distribution

The GEV PDF for the instantaneous wind speed, *U*, is given by

$$p\left(U\backslash k,\mu,\sigma\right)=\frac{1}{\sigma}\left[1+k\left(\frac{U-\mu}{\sigma}\right)\right]^{-1-1/k}\exp\left\{-\left[1+k\left(\frac{U-\mu}{\sigma}\right)\right]^{-1/k}\right\} \tag{4}$$

where k is the "shape factor", Kotz and Nadarajah [11]. In GEV parlance, μ is the "location parameter", and σ is the "scale parameter". The conventional mean is related to these parameters by

$$\bar{U}=\mu-\frac{\sigma}{k}\left[1-\Gamma\left(1-k\right)\right] \tag{5}$$

where Γ is the Gamma function. Similarly the standard deviation is

$$\sigma_u=\sqrt{\mu-\frac{\sigma^2}{k^2}\left[\Gamma\left(1-2k\right)-\Gamma^2\left(1-k\right)\right]} \tag{6}$$

The GEV thus has only one additional adjustable parameter, k, than the Gaussian and so is likely to be more general but not excessively complex. When $k < 0$, the GEV becomes a Weibull, sometimes called GEV Type III, which is the most common PDF used for wind speed analysis. When $k = 0$, (4) becomes

$$p\left(U\backslash 0,\mu,\sigma\right)=\frac{1}{\sigma}\exp\left\{-\left(\frac{U-\mu}{\sigma}\right)-\exp\left[-\left(\frac{U-\mu}{\sigma}\right)\right]\right\} \tag{7}$$

This is the GEV Type I or Gumbel PDF. When $k > 0$, the GEV Type II or Frechet distribution results. In the present work, (4) was applied to the PDFs from the 200 values of *GF* in each bin of 10 min-averaged wind speeds, to determine the three parameters k, μ, and σ, and estimate *GF* for extreme wind speeds.

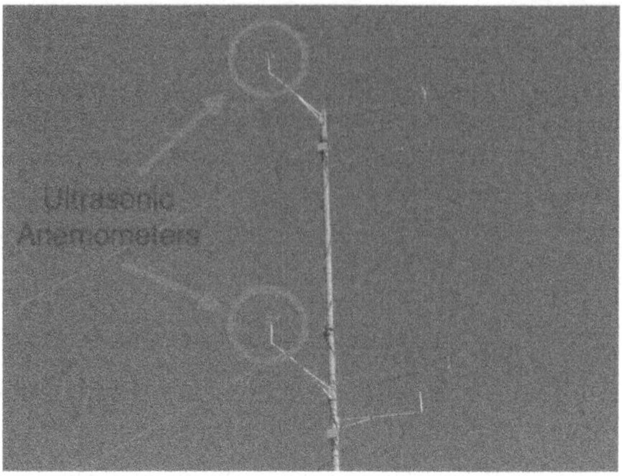

Fig. 1 The top of the 50 m anemometer mast in north Calgary. The two ultrasonic anemometers are on the left hand side of the mast

3 The Measurement Site and Data Processing

The mast is installed in north Calgary (Latitude: 51.17490°, Longitude: −114.22718°) at an altitude of 1273 m above mean sea level and 50 km downwind of the Rocky Mountains. There are no substantial buildings on or near the site, which is grass-covered in summer and usually snow-covered in winter. The fetch in all directions is mainly shallow hills and is more typical of wind farm sites than urban installations, but it was not possible to install the 50 m mast closer to the city centre. The predominant wind direction—and the direction of the strongest winds and highest turbulence—is west-southwest, i.e. from the Rocky Mountains. Using data from the mast, Sherry [12] determined the effect of atmospheric stability on the turbulence structure of the wind and gave more details of the installation and general wind characteristics. Figure 1 shows the top of the mast with the two ultrasonic anemometers. The data described here comes from the top one and no account was taken of atmospheric stability in processing the data.

692 h of data were sampled at 100 Hz. Each hour of data was divided into 10 min bins. Wind speeds in each bin were averaged to give $\bar{U}$, and then the original data was binned into 3 s segments to determine 200 values of *GF* for that period. The data was checked for continuity and consistency before *GF*s were fitted using Eq. (4) to determine the three parameters of the GEV. Given the forms of Eqs. (5) and (6), this fitting is not a trivial process. The details of the data cleaning and the Matlab codes used to fit the PDF can be found in Antonio [3]. This is not a standard use of the GEV which is designed for extreme value statistics, but it can be viewed as a flexible fitting of the data using a reasonable number of parameters. The GEV fits were plotted as a function of 10 min mean wind speed. The *GF* itself is a random variable so it is impossible to specify a unique value.

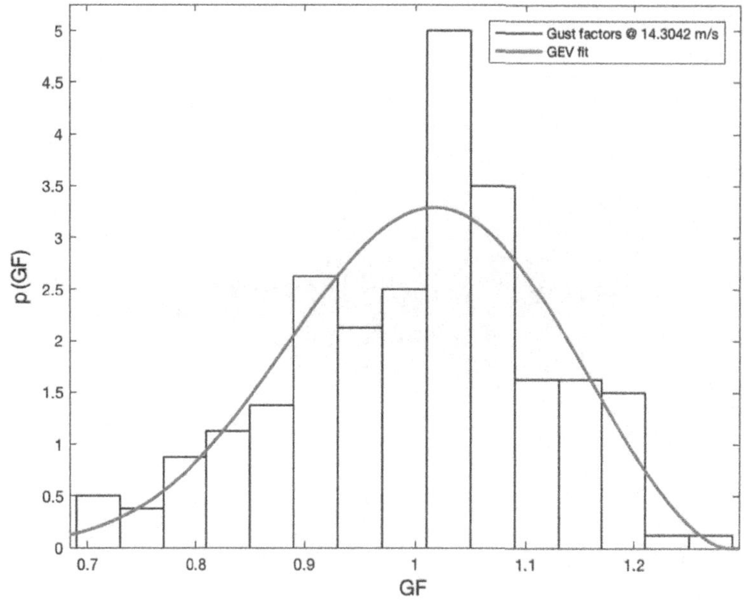

Fig. 2 GEV fitted to the PDF of gust factor. This is a 10 min sample with a mean wind speed of 14.30 m/s

4 Results

The GEV distribution was fitted to 4152 samples of gust factors, each sample with 200 data points. The probability density function of the GEV distribution fitted to a 10 min set of gust factors is plotted in Fig. 2 for the highest mean wind speed, $\bar{U}$, obtained in this study, 14.30 m/s. The maximum gust was 19.03 m/s.

The results of this investigation are based on two methods described below, both of which include the application of the GEV distribution.

The data did not provide high values of $\bar{U}$, but Figs. 3, 4 and 5 are typical in showing that the GEV parameters converged with increasing $\bar{U}$ so that extrapolation to 40 m/s or higher is reasonable. The red lines in these figures show data fits meant to test the high wind speed dependence of the shape, k, location, μ, and scale, σ, parameters. These asymptote to −0.33, 0.96 and 0.13, respectively. Since k is mostly negative, the GEV reduces to the Weibull distribution for most of the datasets.

Those values were used to estimate gust factors using a method described by Antonio [3], which gives a gust factor of 1.26 with a 99% confidence. In this method, a "normalized" GEV distribution of gust factors is generated using the values of k, μ, and σ above, which were found using regression analysis. Then a single value of GF with a probability of 99% is selected from the GEV PDF.

Using another method also described by Antonio [3], we find that 99% of GFs approach a mean value of 1.28. In this method, the GEV distribution is fitted to each

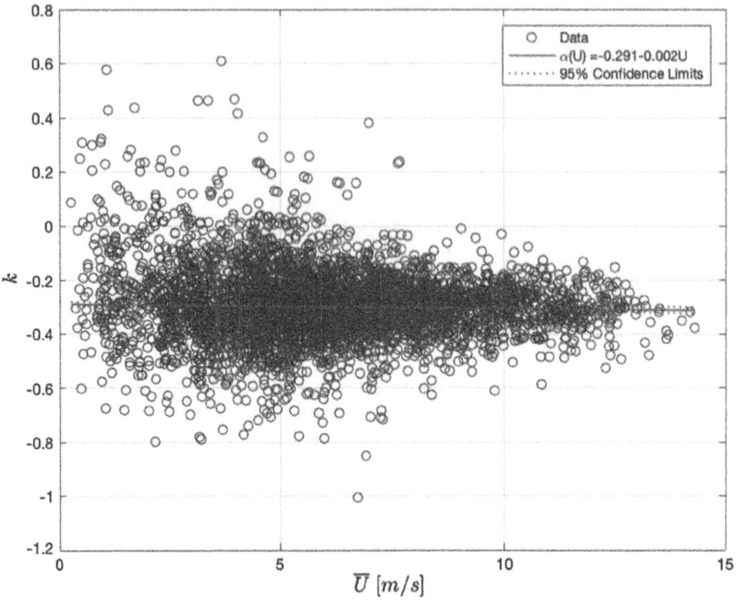

Fig. 3 Shape parameter dependence on wind speed

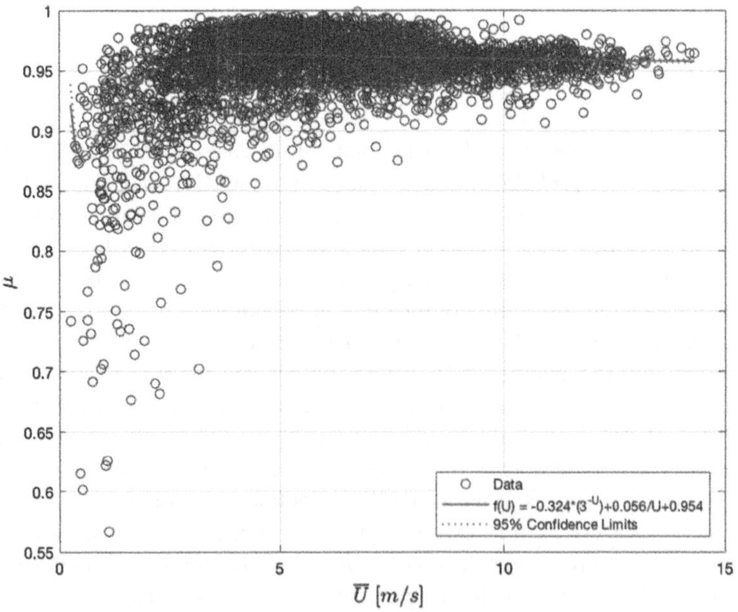

Fig. 4 Location parameter dependence on wind speed

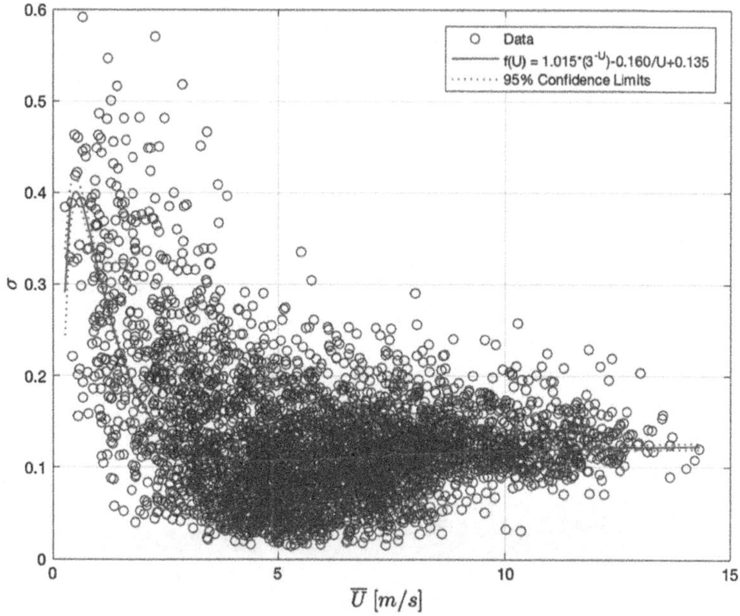

Fig. 5 Scale parameter dependence on wind speed

10 min set of gust factors. Then, all *GFs* with probabilities of 99% are selected from each GEV distribution. These *GFs* are plotted as a function of $\bar{U}$, as shown in Fig. 6. A single value for *GF* is therefore obtained using regression analysis. It is clear from Fig. 6 that *GF* decreases with increasing wind speeds. This trend was also observed by Wang et al. [6] and Davis and Newstein [8]. Figure 6 also shows *GF* as implied by the IEC Standard [1] in the form of Eq. (3). The asymptotic result $GF \rightarrow 1.3$ appears slightly high.

Since the square of the gust factor is used in estimating extreme loads, the values of 1.26 and 1.28 represent a 19% and 16% reduction in these loads compared to the value used in the IEC standard, respectively.

5 Conclusion

This paper describes the determination of the gust factor, defined as the ratio of the 3 s to 10 min average wind speed. 692 h of ultrasonic anemometer data from a 50 m anemometer mast in the semi-urban setting in north Calgary were processed. For each 10 min period, the gust factors were fitted by the Generalized Extreme Value (GEV) probability distribution function. The GEV parameters: shape, location and scale, asymptote to a constant value at high wind speeds, suggesting that a universal

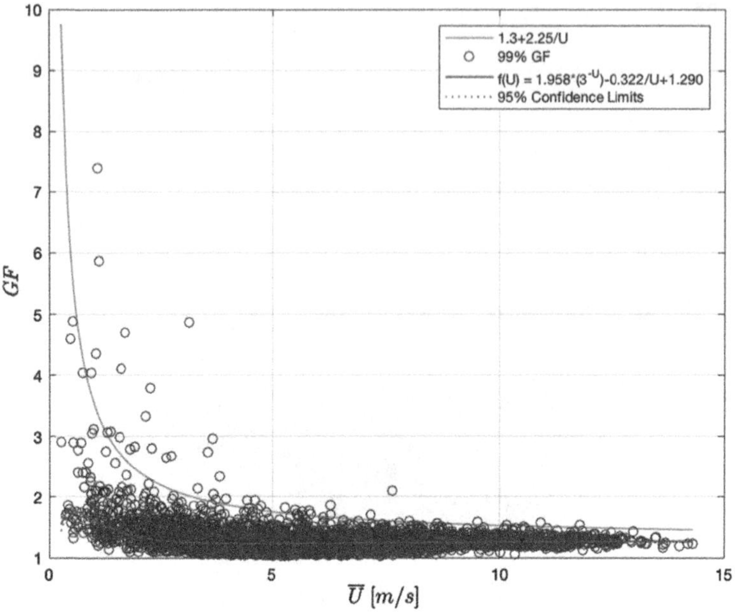

Fig. 6 Gust factors dependence on wind speed

value of the gust factor may be appropriate. It was observed that the gust factor decreased as the wind speed increased.

Gust factors found in this analysis, 1.26 and 1.28, are lower than the gust factor of 1.4 used in the IEC Standard for small wind turbines, [1]. This suggests that [1] is conservative in estimating extreme wind loads by around 19%.

Acknowledgements This work was supported by the Schulich endowment to the University of Calgary. The anemometer mast was provided by WSP, as part of their involvement with an Industrial Research Chair in Renewable Energy funded by NSERC and the ENMAX Corporation.

References

1. IEC: Wind turbines—Part 2: Small wind turbines. IEC Standard 61400-2: Edn 3.0 2013-12 (2013)
2. Holmes, J.D.: Wind Loading of Structures. CRC Press (2015)
3. Antonio, O.L.C.: Extreme wind speed estimation for calgary. M.Sc. thesis, University of Calgary (2017)
4. Davenport, A.G.: Gust loading factors. J. Struct. Div. ASCE **93**, 11–34 (1967)
5. Solari, G., Kareem, A.: On the formulation of ASCE7-95 gust effect factor. J. Wind Eng. Ind. Aerodyn. **77**(78), 673–684 (1998)
6. Wang, X., Chen, B., Sun, D., Wu, Y.: Study on typhoon characteristic based on bridge health monitoring system. Sci. World J. 1–11 (2014)

7. Balderrama, J., Masters, F., Gurley, K.: Peak factor estimation in hurricane surface winds. J. Wind Eng. Ind. Aerodyn. **102**, 1–13 (2012)
8. Davis, F.K., Newstein, H.: The variation of gust factors with mean wind speed and with height. J. Appl. Meteorol. **7**(3), 372–378 (1968)
9. Anastasiades, G., McSharry, P.E.: Extreme value analysis for estimating 50 year return wind speeds from reanalysis data. Wind Energy **17**, 1231–1245 (2014)
10. Ishihara, T., Yamaguchi, A.: Prediction of the extreme wind speed in the mixed climate region by using Monte Carlo simulation and measure-correlate-predict method. Wind Energy **18**(1), 171–186 (2015)
11. Kotz, S., Nadarajah, S.: Extreme Value Distributions—Theory and Applications. Imperial College Press (2000)
12. Sherry, M.: The effect of stability on the intermittent nature of atmospheric winds. In: AIAA 33rd Wind Energy Symposium (2015)

Generation of Uniform Turbulence Profiles in the Wind Tunnel for Urban VAWT Testing

A. Carbó Molina, G. Bartoli and T. De Troyer

Abstract This study presents an innovative wind tunnel approach to evaluate the efficiency of Vertical-Axis Wind Turbines (VAWT) in turbulent flows, to study their integration in urban environments. The first part of the research is devoted to obtaining highly-turbulent wind profiles in the wind tunnel, with the use of different configurations of square grids. A careful study and validation of this technique is done, in order to obtain uniform wind conditions with the adequate values of turbulence intensity and length scales to model the urban flows. The set-up is used to test a H-Darrieus VAWT under values of turbulence over 5%, in comparison with the operation of the turbine under free stream. The preliminary results show that high levels of turbulence do have a significant effect in turbine performance, with increases reaching 15%. The errors in the measurements are analysed to identify the ways to improve the set-up and possible future developments.

Keywords Turbulence · Wind Tunnel · Wind Energy · VAWT

List of Abbreviations

A	Frontal Area (m^2)
b	Bar width (m)
c	Turbine blade chord (m)

A. Carbó Molina (✉) · G. Bartoli
Department of Civil and Environmental Engineering, University of Florence, Via Sta. Marta, 3, 50139 Florence, Italy
e-mail: andreu.carbo.molina@dicea.unifi.it

G. Bartoli
e-mail: gbartoli@dicea.unifi.it

T. De Troyer
Thermo and Fluid Dynamics (FLOW), Vrije Universiteit Brussel, Pleinlaan 2, 1050 Brussels, Belgium
e-mail: tim.de.troyer@vub.ac.be

L. Battisti and M. Ricci (eds.), *Wind Energy Exploitation in Urban Environment*, Green Energy and Technology, https://doi.org/10.1007/978-3-319-74944-0_3

C_P	Power Coefficient (–)
C_{Pfs}	Maximum Power Coefficient at free stream conditions (–)
D	Diameter of the rotor (m)
f	Frequency (Hz)
H	Height of the rotor (m)
I_u	Intensity of Turbulence in wind direction (–)
L_{ux}	Integral Length Scale of turbulence in wind direction (m)
M	Mesh size (m)
Q	Torque (Nm)
R	Turbine Radius (m)
U	Longitudinal wind speed (m/s)
u	Variable component of wind speed in wind direction (m/s)
V	Total wind speed (m/s)
v, w	Wind speed components transversal to wind direction (m/s)
x	Distance from grid (m)
λ	Tip Speed Ratio (–)
λ_{fs}	Optimal Tip Speed Ratio for the free stream case (–)
ρ	Density of air (kg/m^3)
σ_x	Standard deviation of variable x
ω	Angular speed (s^{-1})
HAWT	Horizontal-Axis Wind Turbines
VAWT	Vertical-Axis Wind Turbines

1 Introduction

1.1 Wind Harvesting in Urban Areas

Wind harvesting in urban areas imposes a specific challenge on the energy industry. It could reduce drastically the transportation costs by on-site consumption, and become a key factor for the autonomy of the future smart cities. Moreover, placing wind turbines on the roof of skyscrapers allows to capture the stronger wind at high elevation [1].

However, most of the urban wind installations have turned out to be a complete failure. Still or underperforming turbines also give poor publicity for the technology. An important problem that urban wind turbines face is that the wind climate in cities is very complex. The wind typically has low intensity, high variability, high levels of turbulence, and inclined and even reversed flows.

A careful siting analysis is not usually done, so the turbine is placed in inconvenient locations. Furthermore, the prototypes are mostly designed and tested for smooth winds with low turbulence ($I_u < 1\%$ in a wind tunnel), while inside cities these values easily surpass 10% [2].

1.2 The Importance of Turbulence

Turbulence is said to exert a considerable effect in turbine operation: increase of fatigue, unpredictability of energy production or influence on stall conditions [3]. From the description of an unidirectional flow with 3D turbulence [4], it can be deduced that turbulent flow is more energetic due to the fact that the variable components of the flow *u, v* and *w* (Eq. 1) add up to the power available by the law of the cube (Eq. 2) [5]:

$$|V| = \sqrt{(U+u)^2 + v^2 + w^2} \tag{1}$$

$$P_{av} = \frac{1}{2}\rho A V^3 \tag{2}$$

It is not clear, though, if turbines are able to harvest this extra power, or, if on the other hand, fatigue and aerodynamic disturbances weight too much against operating in turbulent conditions. In HAWTs (Horizontal-Axis Wind Turbines), it is generally searched not to operate in turbulent flows to avoid the turbine being in yaw state [6]. This fact does not occur for VAWTs (Vertical-Axis Wind Turbines) due to their omnidirectionality, which is considered one of their main advantages over HAWTs. Experiences like [5, 7, 8] for open field cases suggest that fact, but to date, there is no study in a controlled environment to assure it.

The goal of the study is to develop a set-up that allows to individualize the effect of turbulence in turbine performance, in the controlled conditions of a wind tunnel. This will not only be used to test VAWT in urban conditions, but also to identify the best-performing designs and improve them for urban operation.

2 Methodology

2.1 Wind Tunnel and Measurement Techniques

The research is performed at the Vrije Universiteit Brussel, in Belgium. The wind tunnel of the Department of Mechanical Engineering is open circuit, of the Boundary Layer type, with a length of 11 m that allow to develop stable turbulence intensity and length scale levels before the test section. The test section, with 1.04 m of height and 2 m width, permits testing scaled buildings and small wind turbines (Fig. 1).

The wind flow characteristics are measured with a Constant Temperature Hot-Wire Anemometer, model Dantec mini-CTA 54T42. Its calibration is done using a Pitot tube. The time histories obtained from the hot-wire are acquired by a National Instruments Card and processed using Labview software to obtain mean speed U and Turbulence intensity I_u, calculated as in Eq. 3:

Fig. 1 Wind tunnel of the Department of Mechanical Engineering of the Vrije Universiteit Brussel, Belgium

Fig. 2 Example of non-dimensionalized turbulence spectra obtained in the measurements (blue), fitted to Von Karman spectrum (red)

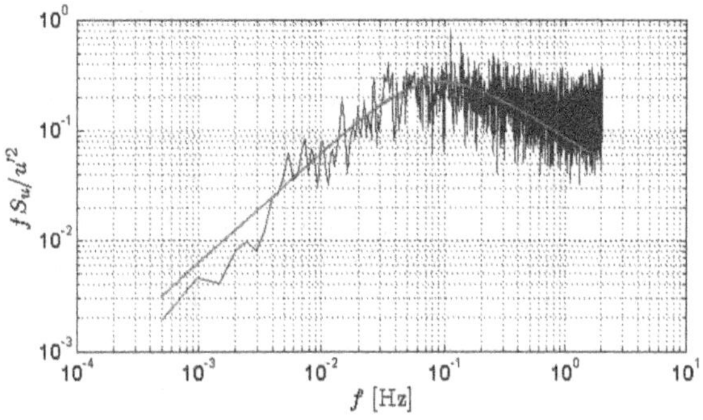

$$I_u = \frac{u_{rms}}{u_{mean}} = \frac{u}{U} \tag{3}$$

where u is the longitudinal component of the wind speed.

The values of the Integral Length scale L_{ux} are calculated according to Taylor's frozen-eddy hypothesis and the results are confirmed by fitting the expression proposed by von Kármán to the measured spectrum [9]. The Von Kármán spectrum gives a good representation of the turbulence inside the wind tunnels and is often used for consistency with analytical expressions [10]. Figure 2 represents the normalized spectrum obtained at the test section centre with the grid at $x = 3.75$ m and $V = 9$ m/s, which is fitted to the formula for the Von Kármán spectrum announced in [10] resulting in a $L_{ux} = 0.11$ m.

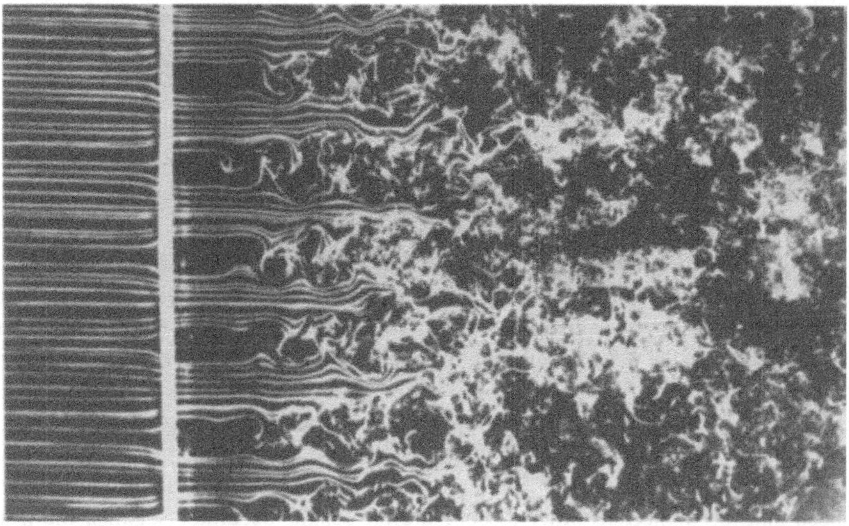

Fig. 3 Smoke visualization of the flow passing through a turbulence grid (photograph by Thomas Corke and Hassan Nagib, from van Dyke [14])

2.2 Squared Grids to Generate Turbulence

Squared turbulence grids have been used for decades to generate artificial turbulence inside wind tunnels. They have been known to provide quasi-isotropic, Reynolds independent profiles, and allow to reach turbulence levels of $I_u > 15\%$ [11]. The turbulence intensity depends on the bar size b, and decreases with the distance x from the grid (turbulence decay). The empirical law proposed by Laneville [12] describes accurately the I_u levels found downstream of the grid (Eq. 4):

$$I_u = 2.58 \left(\frac{x}{b}\right)^{-8/9} \tag{4}$$

The complication falls in recreating the length scales of this turbulence. In urban environments, the turbulence length scales are of the order of $L_{ux} \sim 1$ m [13]. The size of the eddies depends on the bar size b, therefore this supposes a problem for medium-sized wind tunnels, where not-too-wide bars should be used to avoid excessive drag from the grid. An empirical law by Roach [11] for L_{ux} is presented in Eq. 5:

$$L_{ux} = 0.2b \left(\frac{x}{b}\right)^{1/2} \tag{5}$$

It is true that those turbulent structures increase in length downstream, but also here there is a limitation due to the decay of turbulence intensity. Figure 3 shows a smoke visualization of laminar flow passing across a turbulence grid [14].

Table 1 Characteristics of the first grid build for the study

Mesh size M	0.33 m	b/M	0.22
Bar width b	0.072 m	C_D	1.04
Distance from grid x	3.75 m	x/M	11.4
Predicted I_u [12]	7.7%	Predicted L_{ux} [11]	10.4 cm

After a careful study of literature and previous studies at the University of Florence [9, 12, 15, 16] it is clear that the largest scales that could be generated would be at the order of $L_{ux} \sim 0.1$ m, by using large mesh sizes *M*. This way, they would still be larger than the chord of the wind turbine blade, which is $c = 0.05$ m. According to Bearman and Morel [15], when $L_{ux} \gg c$, the gross effect of turbulence can be estimated using a quasi-steady assumption. Though not exact, this assumption is considered as valid in this case.

The last feature that is asked from the grid is to create an uniform profile of wind speed and turbulence, in order to compare the results accurately against the free stream conditions. According to literature, the flow will be fully developed after a distance $x/M > 5$. However, when using mesh sizes larger than 10% of the wind tunnel dimension, the wind profile is heavily unstable [15], and small changes of geometry can lead to significant modifications in the flow. This is a condition that cannot be fulfilled and at the same time maintain the high values of L_{ux}. Therefore, a careful parametric study has been done to "tune" the grid parameters (Mesh size *M* and bar size *b*), to obtain the desired uniform profiles.

2.3 Parametric Study to Achieve Uniform Turbulence Profiles

The starting point is a regular mesh of $M = 0.33$ m, with 6 cells horizontally and 3 vertically (wind tunnel test section is 2×1.04 m). The width of the bar is limited by the blockage it causes in the wind tunnel. From the thesis of Laneville [12] an expression for the drag coefficient C_D is drawn (Eq. 6):

$$C_D = \frac{\frac{b}{M}\left(2 - \frac{b}{M}\right)}{\left(1 - \frac{b}{M}\right)^4} \tag{6}$$

It can be seen that C_D increases drastically for values of $b/M > 0.25$. Therefore, a width slightly under this value is chosen. The grid is placed at around 4 m from the test section to obtain values of moderate turbulence. The characteristics of the grid are shown in Table 1, and the grid is depicted in Fig. 4.

For simplicity, the wind and turbulence profiles are only analysed in the vertical direction, which is the most challenging to obtain uniform profiles, as the size of the mesh is much larger than 10% of the wind tunnel height.

Fig. 4 Uniform grid with 6 × 3 cells

Fig. 5 Wind speed and turbulence intensity profiles obtained with free stream and Grid 1 at $x = 3.75$ m

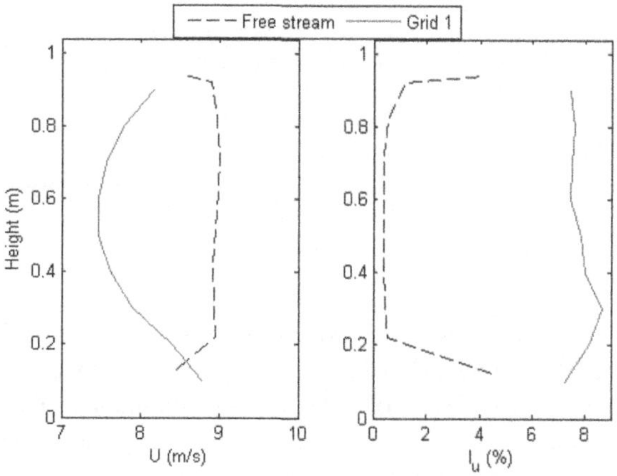

Figure 5 shows the profiles of wind speed and turbulence created by the first grid, compared with the measurements taken with a clear wind tunnel. It can be seen that the profile obtained with the grid cannot be considered uniform at all, as the flow is clearly accelerated next to the walls. While the free stream velocity has a $\sigma_u = 0.05$ m/s (excluding boundary layer), this value for the Grid 1 profile is $\sigma_u = 0.41$ m/s, clearly unacceptable for comparison. This can be caused by the fact that the 2 central bars create larger vortexes than the ones in the walls (immersed in the boundary layer). This creates a bluff body effect in the wake of the grid.

The good part of these results is that the turbulence profile is reasonably stable and its average value is $I_u = 7.89\%$, not far from the one predicted by literature.

The solution to this was to reduce the space between the bars and the walls, to stop this acceleration. The grid, then, was modified several times, varying the mesh size M to select the best gap next to the walls (Fig. 6).

Fig. 6 Different modifications of the grid to obtain uniform flow. From left to right, grids 2 ($M = 0.33$ m), 3 ($M = 0.35$ m) and 4 ($M = 0.40$ m)

Fig. 7 Normalized profiles obtained with the different modifications of the grid at $x = 3.75$ m

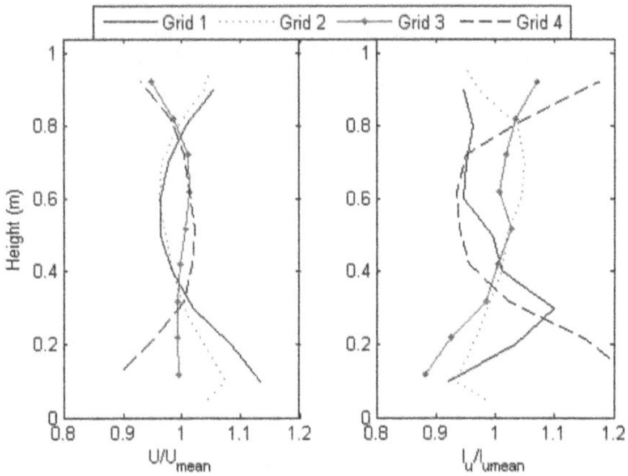

For an easier comparison, the results are normalized with the average values (excluding boundary layer), as the study searched the uniformity of the profile more than precise levels of wind speed or turbulence. Therefore, it was predicted by literature, and will be proven by upcoming results, that the turbulence generated by squared-bars grids is no speed dependant.

Figure 7 shows the comparison between the normalized profiles of wind speed and turbulence intensity obtained with the different grids. It can be seen that in Grid 2, reducing by half the gap to the wall, the accelerated flow is still found. On the other side, with Grid 4 the smallest gap near the walls causes the opposite effect, a jet. Grid 3 seems to be the intermediate point, although, as predicted in the literature, obtaining a uniform profile is highly complicated for these mesh sizes.

In fact, from a more extensive study it has been proved that small changes in geometry (smaller than 1 cm) have large impact on the flow, and even a careful construction of the grid leads to asymmetries in the flow that cannot be easily solved. The conclusion is that an optimal flow is nearly impossible to achieve in these conditions, so the best approximation to the free stream flow must be chosen, to be able to compare successfully the turbine ratings in the various cases.

Table 2 summarizes the results of the preliminary study, compared with the free stream. The bar size b and the distance to the grid x are fixed. Analysing the standard deviation of the profiles, it can be easily seen that the profile obtained with Grid 3 is the only one that can be compared with the free stream values. Even though the

Table 2 Summary of the parametric grid study

Grid number	Mesh size M (m)	b/M	Gap next to walls (m)	gap/M	Deviation σ_u (m/s)	I_u *mean* (%)	*Deviation* σ_{Iu} (%)
Free stream	–	–	–	–	**0.05**	0.5	**0.05**
Grid 1	0.33	0.22	0.33	1	0.41	7.89	0.44
Grid 2	0.33	0.22	0.17	0.5	0.23	7.69	**0.24**
Grid 3	0.35	0.21	0.15	0.43	**0.09**	7.75	0.28
Grid 4	0.40	0.18	0.1	0.25	0.21	9.34	0.74

turbulence profile is not as good as the obtained with Grid 2, the uncertainty in wind speed is the main factor to consider when evaluating turbine ratings, as it will be demonstrated in the last chapter.

2.4 *Generation of Different Levels of Turbulence*

In order to test different levels of turbulence, the grid should be approached to the test section. There is a limit in which the flow would still not be developed, which is predicted at $x/M < 5$ by Bearman and Morel [15]. Grid 3 is then moved until 2 m from the test section, which, according to the literature, would provide a $I_u = 13.4\%$ and a $L_{ux} = 0.076$ m. The power of the wind tunnel had to be increased with the rise of turbulence, to obtain the most similar wind speed profiles that was possible. Figure 8 presents the profiles of Wind Speed, Turbulence Intensity and Length Scale obtained with Grid 3 for 9 m/s.

As it can be seen in Fig. 8, the wind speed profile when the grid is closer is not as stable, showing slight acceleration near the walls. This is because of the divergence of the wind tunnel, that has 1 cm more of height at $x = 2$ m than at $x = 3.75$ m. However, the wind speed profiles are relatively uniform and similar in the area of study (outside the boundary layer). About the I_u profile, it can be seen that as the turbulence increase is more difficult to obtain an stable profile. The L_{ux} values are, as expected, lower when the grid is closer, but always higher than $c = 0.05$ m.

The tests were also performed at 8 m/s to study the effect of wind speed. Those two speeds were chosen as there are within the operational range of the turbine and no resonance is found around the optimal performance point. Table 3 summarises the conditions under which the VAWT is tested in the wind tunnel. It can be seen that the change in wind speed has practically no effect in the characteristics of the turbulence, and how the grid in the position $x = 2$ m is not as optimized as before.

In this position, as explained in the previous section, turbulence is significantly higher, and L_{ux} lower (but still over c). The standard deviation of the wind speed and I_u are higher than before, but still under 5% of the main signal value. This drop

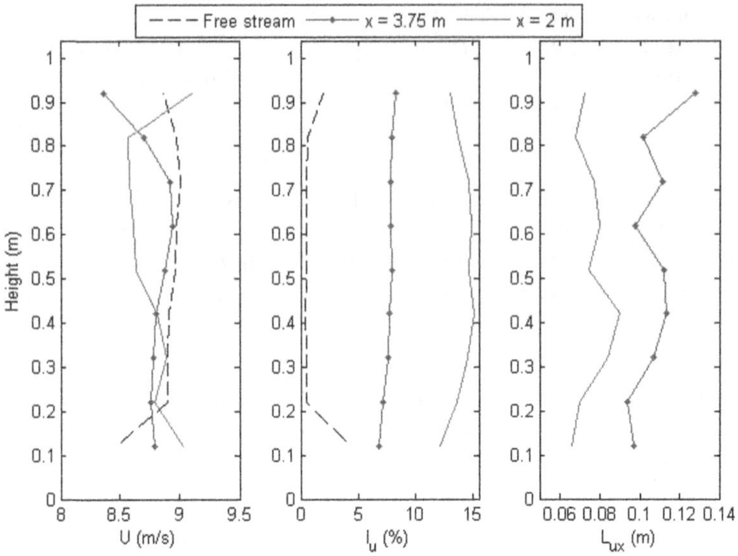

Fig. 8 Profiles obtained with grid 3 at different positions

Table 3 Input conditions for every test done on the VAWT

Grid position *x* (m)	*x/M*	U_{mean} (m/s)	σ_u (m/s)	I_u *mean* (%)	σ_{Iu} (%)	L_{ux} mean (m)	σ_{Lux} (m)
No grid	–	7.88	0.04	0.51	0.06	–	–
No grid	–	8.86	0.05	0.50	0.05	–	–
3.75	10.7	7.90	0.09	7.71	0.28	0.104	0.011
3.75	10.7	8.82	0.09	7.75	0.28	0.105	0.011
2.00	5.7	7.85	0.12	14.34	0.61	0.078	0.008
2.00	5.7	8.76	0.13	14.46	0.55	0.078	0.008

in uniformity of the profiles is due to the fact that the grid has been optimized for $x = 3.75$ m, and when bringing it closer to the test section, the wind tunnel height slightly increases, due to its divergence to avoid blockage. Even though the height difference between the two *x* positions is lower than 1 cm, this is already significant for the flow. Furthermore, at $x = 2$ m the flow is not as developed as in the more distant position.

Fig. 9 VAWT turbine placed in the wind tunnel, with the turbulence grid 3.75 m upstream

2.5 Testing of the VAWT

The VAWT is a 2-bladed H-Darrieus design with $D = 50$ cm and $H = 80$ cm, and it was built in lightweight carbon fibre at VUB (Fig. 9). The blades' profile is a NACA0018 with $c = 5$ cm chord. The turbine is supported by an aluminium modular frame already tested in other VAWT measurements. The torque is measured by Lorenz Messtechnik DR-3000 torque meter (max 5 Nm, precision 0.1% of 2 Nm), and the rotational speed is controlled using variable resistors [17]. The axis of the turbine is coupled to a brushed DC motor/generator, that allows the start-up of the turbine.

The parameter to calculate the performance of the turbine will be the Power Coefficient (C_P):

$$C_P = \frac{P_{turbine}}{P_{wind}} = \frac{Q\omega}{\frac{1}{2}\rho D H V^3} \tag{7}$$

The conventional way of representing C_P in wind turbine studies is plotting it against the Tip-Speed Ratio of the turbine, a non-dimensional number that relates the speed of the blade tip with the incident wind:

$$\lambda = \frac{\omega R}{V} \tag{8}$$

Table 4 Conditions assumed in the turbine combining experiment and literature data [11, 12]

Grid position x from turbine (m)	x/M	I_u *mean* (%)	L_{ux} mean (m)
No grid	–	0.51	–
4.75	13.6	6.2	0.114
3	8.6	9.4	0.090

3 Results

This chapter is divided in two parts; in the first, the experimental results obtained for the test matrix shown on Table 3 are presented. In those power curves error bars are included to have an overview of the accuracy of the method, which was the main objective of the study. The second part of the chapter explains in detail how this uncertainties are calculated, and which are the key factors that influence the method accuracy.

3.1 Power Curves in Different Turbulence Conditions

First of all, it is important to notice that the profiles obtained with the anemometer are not identical to the ones the turbine will see. It can be seen in Fig. 9, that the Hot Wire is placed at the beginning of the test section (top of the image), while the turbine is in its centre (1.00 m downstream). This difference will not affect the shape of the profile (at must, it will be more uniform as it is farther away from the grid). Assuming the turbulence keeps following the laws from Laneville [12] and Roach [11] (Eqs. 2 and 3), the values of turbulence and length scales are fixed as it follows (Table 4).

Independent measures have been taken to ensure the previous assumption, and that the wind profile behind the grid is not modified by the rotation of the turbine. The blockage factor, however, would affect the measurements as the turbine is large compared with the test section ($A_{turbine}/A_{test\ section} = 19\%$). Jeong et al. [18] detected large influence of blockage for A ratio larger than 15%, and according to the correction provided for Pope and Harper [19] the apparent increase of the wind speed for this case would be a non-neglectable 5%, overpredicting the turbine performance. However, in this study the overall performance of the turbine is not the main objective, but the development of the set-up and the influence of turbulence at a constant wind speed. Therefore, no blockage correction will be applied, although to completely validate the methodology, the set-up must be replicated in the future in a lower blockage wind tunnel to assess if the influence of turbulence on the turbine performance is affected by blockage.

To obtain a complete C_P-λ curve, the rotational speed of the turbine is tuned onto different values, to sweep over the various λ. A point of the C_P curve is registered

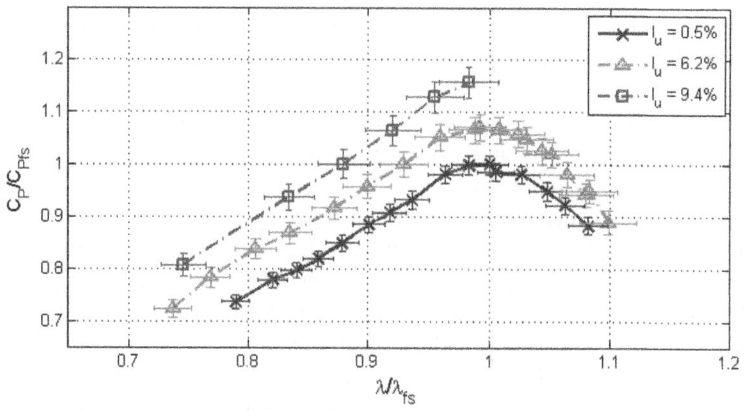

Fig. 10 Power curves for $V = 8$ m/s, non-dimensionalyzed with the optimal values for the free stream case

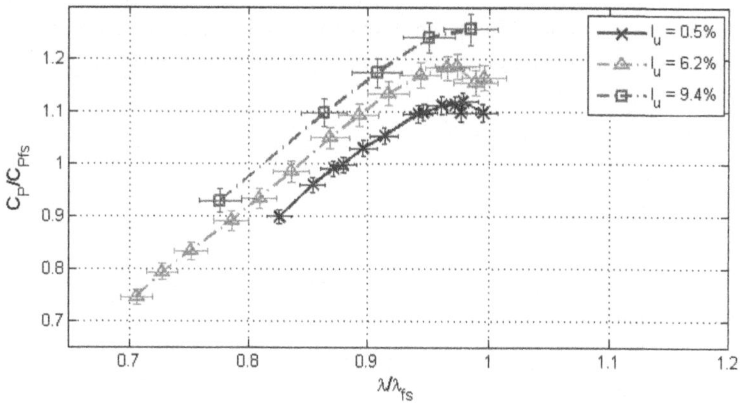

Fig. 11 Power curves for $V = 9$ m/s, non-dimensionalyzed with the optimal values for the free stream case at 8 m/s

when the ω value is constant (± 0.2 Hz) during 1 min. The torque values present huge deviations, due to the large variations of the instantaneous torque along a turbine rotation [3]. However, when averaged during 60 s, the mean value of this torque is accurately determined.

Figures 10 and 11 present the C_P curves obtained for the tests at 8 and 9 m/s. Due to confidential reasons, the C_P and λ values have been non-dimensionalized with the optimum values for the curves at free stream at 8 m/s. It can be seen that the power extracted by the turbine increases significantly with the turbulent flow. As it was stated in the introduction, the turbulent flow is more energetic, and here it can be clearly observed that the VAWT is able to capture that excess of energy.

It can be seen that for moderate turbulence there are many points, and the curve is more defined. This is because the control of the turbine, which is done manually, is much easier in that condition. In free stream the flow was not energetic enough to withstand low λ, while for high turbulence the turbine was much more unstable in its operation.

Comparing the two wind speeds, the Reynolds effect can be appreciated, as in Fig. 11 ($V = 9$ m/s) all the curves are shifted upwards. It also catches the attention that there are no points after C_{Pmax} in the case of $V = 9$ m/s. This is not due to the wind speed, but to the fact that the turbine structure enters resonance around 20 Hz, a fact that can only be counteracted by changing the shaft and all the turbine structure.

Another interesting fact from the graphs regards the error bars, which are larger as turbulence increases. This is due to the fact that the main factor inducing errors is the accuracy of the wind speed, and as it was seen in the previous chapter, the wind profiles for the cases of high turbulence where not as accurate as the free stream ones. These differences in the error bars are described in detail in the next section.

Overall, the increase of C_{Pmax} between the free stream case ($I_u = 0.5\%$), and the highest turbulence ($I_u = 9.6\%$) is about 15%, while the moderate turbulence case ($I_u = 6.2\%$) represents an intermediate result. More values of turbulence should be measured to draw a solid conclusion about the trend of this increase. There is no significant effect of turbulence in the optimal λ, which is, at most reduced slightly with the increase of C_P.

3.2 Uncertainties Calculation

The uncertainties are calculated by the error propagation method, enounced in Eq. 9 [20] and developed in detail in [21]:

$$u_R = \sqrt{\sum_i^N \left(\frac{\partial R}{\partial x_i} u_{xi}\right)^2} \tag{9}$$

To calculate the error in C_P, the variables present in Eq. 7 are analysed:

- Torque Q: the sensor accuracy is the main source of error. The random error is reduced as many samples are taken for each Q.
- Angular speed ω: Same as in torque.
- Density ρ: Density is calculated with the perfect gases law from the values of Pressure and Temperature. The accuracy in those measurements must be considered.
- Geometry D, H: If the measure is done accurately, their contribution in the error is small.
- Wind speed V: The first source of error in V comes from the anemometer calibration. A Hot Wire Anemometer is very precise, but the pitot tube used to do the calibration is not so much, adding uncertainties to the measurement. The second source is the uniformity of the wind speed profile over the turbine swept area (called σ_u in the Methodology section).

Table 5 Contribution of the different parameters onto the C_P uncertainty, for different values of turbulence at the optimal performance point for $V = 8$ m/s

Parameter	Contribution to C_P uncertainty		
	$I_u = 0.5\%$	$I_u = 6.2\%$	$I_u = 9.4\%$
Q	2,12E−03	2,11E−03	2,21E−03
ω	4,37E−04	4,36E−04	4,56E−04
ρ	5,70E−05	4,74E−05	1,67E−04
D	6,88E−04	6,87E−04	7,18E−04
H	1,10E−03	1,10E−03	1,15E−03
V	3,98E−03	5,78E−03	7,08E−03
Total C_P uncertainty	4,71E−03	6,30E−03	7,55E−03

λ error is calculated the same way. Table 5 shows how the different parameters in Eq. 7 contribute to the uncertainty value of C_P in free stream and turbulent cases, for $V = 8$ m/s.

As it was already seen in Figs. 10 and 11, errors increase significantly with the level of turbulence, and it is basically due to the uncertainty in wind speed, as all the other factors remain quite stable.

4 Conclusions

- There is no experiment in literature that proves in a controlled environment that VAWT operate better in turbulent flows.
- The recreation of urban turbulent flows inside a medium-sized wind tunnel is possible regarding I_u levels, while L_{ux} can only be approximated.
- The values of Lux and I_u are successfully predicted by literature.
- Main complication is that when $M/x > 0.1$ the profile is unstable, highly depending on small changes in grid geometry.
- It is possible to obtain uniform profiles by leaving 0.43 M grid spaces next to the walls.
- The performance of the H-Darrieus VAWT increases steadily with turbulence, until a 15% of C_{Pmax} increase for a 10% turbulence.
- The non-uniformity of the wind speed profile is the main source of error in the measurements, even though anemometer calibration and torque sensor accuracy have also significant influence.

4.1 Future Developments

- With the same set-up:
  - Reduce uncertainties.
  - More I_u values to define a trend $P - I_u$.
  - Measures of the effect of turbulence on vibrations and control.
- Modifying set-up:
  - Effect of L_{ux} (by building different grids)
  - Effect of blockage (larger wind tunnels)
  - Effect of turbulence on other turbines.
- Same set-up can be used for HAWT, buildings, bridges, etc.

Acknowledgements The present research takes part in the European Innovative Training Network (ITN) AEOLUS4FUTURE "Efficient Harvesting of the Wind Energy". The project is funded by the Horizon 2020 research and innovation program under the Marie Skłodowska-Curie grant agreement No. 643167.

References

1. Barlow, J.F., Drew, D.R.: Wind flow in the urban environment. In: WINERCOST Workshop 'Trends and Challenges for Wind Energy Harvesting' 'Trends and Challenges for Wind Energy Harvesting', Coimbra, Portugal (2015)
2. Janajreha, I., Su, L., Alan, F.: Wind energy assessment: Masdar City case study. Renew. Energy **52**, 8–15 (2013)
3. Paraschivoiu, I.: Wind Turbine Design: With Emphasis on Darrieus Concept. Presses inter Polytechnique (2002)
4. Olivari, D., Benocci, C.: An Introduction to the Mechanics of Turbulence. Course note 157, Von Karman Institute for Fluid Dynamics (2013)
5. Möllerström, E., Ottermo, F., Goude, A., Eriksson, S., Hylander, J., Bernhoff, H.: Turbulence influence on wind energy extraction for a medium size vertical axis wind turbine. Wind Energy **19**, 1963–1973 (2016)
6. Lubitz, W.D.: Impact of ambient turbulence on performance of a small wind turbine. Renew. Energy **61**, 69–73 (2014)
7. Pagnini, L.C., Burlando, M., Repetto, M.P.: Experimental power curve of small-size wind turbines in turbulent urban environment. Appl. Energy **154**, 112–121 (2015)
8. Bertényi, T., Wickins, C., McIntosh, S.: Enhanced energy capture through gust-tracking in the urban wind environment. In: 48th AIAA Aerospace Sciences Meeting Including the New Horizons Forum and Aerospace Exposition, 4–7 January 2010, Orlando, Florida (2010)
9. Mannini, C., Marra, A.M., Pigolotti, L., Bartoli, G.: Unsteady pressure and wake characteristics of a benchmark rectangular section in smooth and turbulent flow. In: 14th International Conference on Wind Engineering, Porto Alegre, Brazil (2015)
10. Burton, T.: Wind Energy Handbook. Wiley (2001)
11. Roach, P.E.: The generation of nearly isotropic turbulence by means of grids. Int. J. Heat Fluid Flow **8**(2), 82–92 (1987)

12. Laneville, A.: Effects of turbulence on wind induced vibrations of bluff cylinders (Ph.D. thesis). University of British Columbia, Vancouver, Canada (1973)
13. Dallman, A.R.: Flow and turbulence in urban areas (PhD thesis). University of Notre Dame, Indiana, USA (2013)
14. Van Dyke, M.: An Album of Fluid Motion, 14th ed. Parabolic Press, Inc. (1982)
15. Bearman, P.W., Morel, T.: Effect of free stream turbulence on the flow around bluff bodies. Prog. Aerosp. Sci. **20**(2), 97–123 (1983)
16. Baines, W.D., Peterson, E.G.: An investigation of flow through screens. Trans. ASME **72** (1951)
17. De Troyer, T., Runacres, M.: Wind Tunnel Testing of a Pair of VAWTs for Offshore Applications (2016)
18. Jeongl, H., Lee, S., Kwon, S.D.: Wind tunnel interference effects on power performance of small Darrieus wind turbines. In: The 2014 World Congress on Advances in Civil, Environmental, and Materials Research (ACEM14), Busan, Korea (2014)
19. Pope, A., Harper, J.J.: Low Speed Wind Tunnel Testing. Wiley, New York (1966)
20. Dénos, R.: Fundamentals of Data Acquisition and Processing. Course Note 171, von Karman Institute for Fluid Dynamics (2005)
21. ITTC—Recommended Procedures and Guidelines, Uncertainty Analysis: Particle Imaging Velocimetry. In: 25th International Towing Tank Conference (2008)

12. [illegible]
13. [illegible]
14. [illegible]
15. [illegible]
16. [illegible]
17. [illegible]
18. [illegible]
19. [illegible]
20. [illegible]

Operational Behaviour of SWT

M. Peppoloni, A. Hirschl and K. Leonhartsberger

Abstract Alongside photovoltaics, small scale wind turbines (SWT) represent one of the few possibilities to yield energy in an environment friendly way in urban areas and thus a step towards "nearly zero energy"-buildings. However, due to a lack of experience in this regards, in many cases the impact of the environment on the SWT's performance has been neglect. A poor energy yield, frequent faults and failures are the result [1]. On top of this, safety aspects as well as the repercussions on the building and the people living in it, need to be considered. Therefore, the objective of the presented work is to examine three different roof mounted SWT technologies (Savonius, Derrieus-Helix and three blade horizontal rotor) in simulation environment as well as under real life conditions while considering the influence of the roof shape. The main focus thereby is: A validation of currently available simulation models. An evaluation of the impact of complex obstacles on the local flow pattern as well as on the SWT under real working and environment conditions. A metrological evaluation of the repercussions of roof mounted SWT on its performance as well as on the building, the resident people and the direct environment in terms of noise, infra-sound, body-borne sound, vibrations and safety relevant aspects. In order to meet this target, two buildings (one with flat roof and one with gabled roof) are built in the "Lichtenegg" energy research park. While previous studies relied on simulations and wind tunnel tests on scaled models to gain understanding for the undisturbed operation of SWT [2, 3], the test infrastructure deployed in the presented work allows monitoring a SWT installed on the roof of a building under real life conditions. In order to get the full picture of the operational behaviour of the tested SWTs, the following parameters are measured: Inflow conditions at several points around the building. Energy yield and selected operating parameters of the SWT (e.g. power, RPM, …). Vibration on the SWT and on/in the building. Noise, infra-sound and body-borne sound. Since vibrations have caused issues on building mounted SWT in several occasions [1], the efficiency of vibration decoupling elements are evaluated. This is done by measuring the vibrations as well as structure-borne sound in the building over a measurement

M. Peppoloni (✉) · A. Hirschl · K. Leonhartsberger
Department for Renewable Energy, University of Applied Science Technikum Wien, 1210 Vienna, Austria
e-mail: mauro.peppoloni@technikum-wien.at

L. Battisti and M. Ricci (eds.), *Wind Energy Exploitation in Urban Environment*, Green Energy and Technology, https://doi.org/10.1007/978-3-319-74944-0_4

period of 3 months and comparing the results with those obtained with the custom fitted decoupling elements installed. Based on this comparison, recommendations regarding the use of vibration decoupling elements for building mounted small wind turbines are given. In regard to an estimation of the social and ecological impact of SWT, a technological impact assessment of the three deployed models is performed. Besides the Global Warming Potential, other factors such as energetic and financial amortisation periods are determined.

Keywords Small wind turbine · Building mounted · Turbulence · Vibration Oscillation · Acoustic noise · Measurement

1 Scope of Work

The main goal of the presented work is to gain practical understanding for the airflow around low rise building as well as for the operation of building mounted SWT and the related challenges and issues. Increased turbulence intensity and inhomogeneous wind speed distribution over the rotor swept area cause losses and increased oscillations [4]. While most previous R&D activities relied only on CFD simulations and scaled model testing in wind tunnels [5] to investigate the airflow around buildings and through SWT rotors, the project "SmallWindPower@Home" enhances the CFD simulations by validating them through real life measurement on buildings in an open field test site.

In a first step, CFD simulations of air flow around low rise buildings with various roof shapes are performed using common simulation models. This is done to help gaining applied knowledge on turbulence conditions, wake and stall effects, bubble separation characteristics, acceleration of the air flow and other aerodynamic effects [6, 7] over and around the specific buildings deployed in this work.

In a second step, the two one-storied buildings—one with flat roof and one with gabled roof—are constructed in the "Lichtenegg" research park and wind conditions are measured on several locations around the buildings by means of 3D ultrasonic anemometers. The measurement results are then compared with the predictions of the CFD simulations in order to validate or to improve the models if needed.

In a next step, three different SWT types are tested in sequence, each mounted on a free standing mast, on the flat roofed and the gabled roof building. Each combination of turbine and supporting structure is operated and measured for at least three months. During the long term evaluation periods, the turbine's operation data is recorded. The power curve and normalized energy yield measured of each combination is then compared to identify whether the influence of the building is rather positive or negative on the turbines performance and which type of turbine is better suited for the altered urban-like wind conditions. Additionally, oscillation measurements on the tower and in the building are performed in order estimate the occurring amount of force which may cause excessive wear of the structure and eventually lead to failure [8]. Furthermore, these measurement results are supposed to quantify the damping characteristics of the deployed oscillation decoupling element, the gabled roof structure and the building itself. The oscillation measurement combined with other parameters like wind speed and direction, rpm and power of the turbine help to gain understanding for interaction of the turbine with the urban-like wind conditions.

As described by Mollasalehi et al. [9], acoustic emissions caused by rotating and vibrating parts of the turbine or aero-acoustics effects [10] can be amplified by the tower. In order to assess the sound levels human beings in the building are exposed to, acoustic sensors at several locations in the building measure the sound and infrasound emissions. The structure borne sound and vibrations are recorded in the building by means of human vibration sensors and compared with maximum tolerable values according to EU-directive 2002/44/EG.

Based on these findings, a socio-economic assessment is performed to determine the social impact in a scenario where SWTs are deployed in large numbers in urban or pre-urban areas. This assessment takes into consideration additional aspects like shadow flickering, aesthetical appearance, ice throw and other safety related aspects which can have repercussions on nearby inhabitants quality of life.

Finally, a life cycle assessment is performed to rate the overall CO_2 foot print and flow of material as well as energetic and economic amortisation periods of the deployed turbines.

2 Test Site

Situated on the ridge of a hill at 800 m above sea level and featuring an average annual wind speed above 5 m/s, the "Lichtenegg" energy research park serves as test field for small wind turbines since 2013. According to [11], these conditions paired with fact that no significant obstructions are located within 200 m, make this site an excellent choice for performing long term operation evaluation of SWT. The outstanding wind resource is confirmed by the meteorological measurements that have been performed over the last few years.

The meteorological analysis has shown an average annual wind speed of 5.2 m/s measured at 16.5 m above the ground as seen in Fig. 1. The most frequent wind directions are north-northwest (NNW) with a frequency of occurrence of 23% and south (S) with a frequency of occurrence of 21%. For these characteristics, the "Lichtenegg" energy research park was chosen as a test site for the empirical investigations and measurements in the project "SmallWindPower@Home".

Figure 2 shows the picture of the site (left) side and a site plan (right) of the entire energy research park. Location number 18 in the site plan indicates the position of the wind measurement mast in the east of the research park. Several wind speed and wind direction sensors are installed at 19, 16.5, 15 and 10 m above the ground. The wind speed and direction is measured with 3D ultra-sonic anemometers at 16.5 and 19 m above the ground to acquire data in all three dimensions and gather statistical data of the annual wind direction and wind speed distribution. At 10 m above ground,

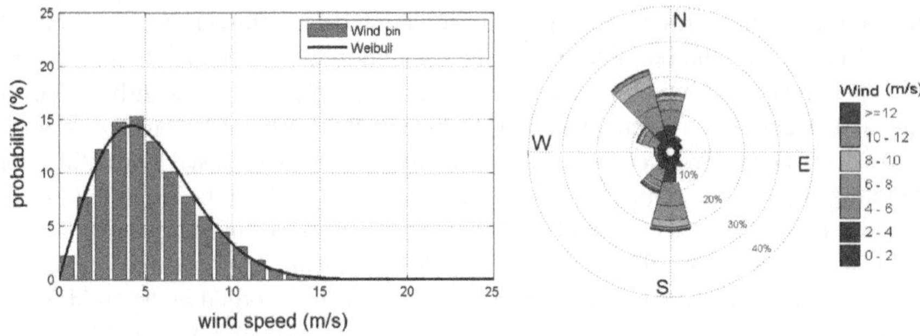

Fig. 1 Weibull wind profile and wind rose

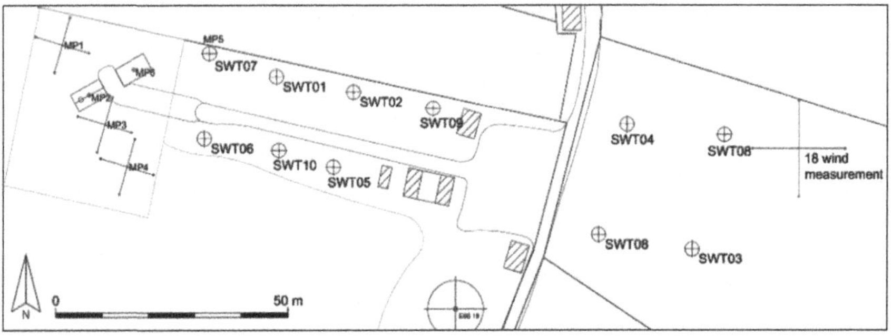

Fig. 2 Energy research park Lichtenegg and site map

the wind speed and wind direction is measured by means of classic cup anemometers and wind vanes.

3 Test Infrastructure

Based on the wind direction distribution, the location and orientation of the new test infrastructure was determined. The site map in Fig. 2 (left) shows the selected area on the western side of the research park including the two buildings and met masts marked as MP1 to MP6. The dimensions of the new test site are 38 m in length and 32 in width.

Figure 3 shows the measurement points MP1 to MP4 which are in line with the main wind direction coming from NNW (331°). The first mast (MP1) is installed in a distance of 10 m from the building. To evaluate the impact of the altered flow pattern for the SWT, the wind conditions at turbine hub height above the building is measured at MP2. The flow pattern and wake effects behind the building are captured in 5 m distance from the building edge at MP3 and in 15 m distance from the building at MP4. On each mast, tree 3D ultra-sonic anemometers are installed at a height of

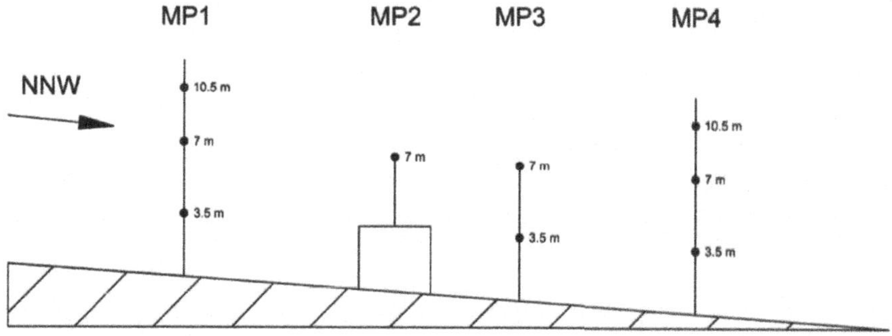

Fig. 3 Location of wind sensors

3.5, 7.5 and 10.5 m above the ground. Additional measurements at hub height on the second building (MP6) and on the free standing mast SWT07 (MP5) enable a site calibration which is performed before installing the turbines on the buildings.

The buildings are set up in an orthogonal orientation compared to the measurement line. This setup allows to record the flow condition for situations where the wind meets the edge of the building in a right angle in case of NNW wind direction as well as for situations where the wind meets the edge of the buildings in a steeper angle in case of south wind. In both situations the met mast situated in upwind direction allows measuring the undisturbed incoming flow pattern.

A site calibration performed before the construction of the two buildings allows comparing the flow pattern around the buildings with the flow pattern in an open field situation with identical slope and boundary conditions.

Figure 3 shows the alignment of the met masts along the measurement axis including the height of the anemometers. As the terrain is sloping downwards with an inclination of 10° and two almost opposite wind directions are evaluated, a potential influence of the terrain inclination can be evaluated.

Studies with wind tunnel tests have shown that different roof shapes lead to a different impact on the flow pattern influencing the performance of a roof-top installed small wind turbine [12]. In order to assess the impact on the turbines performance, two typical roof shapes, gabled and flat roof are evaluated during die tests. The two buildings are realized as precast concrete garages to reproduce a typical application scenario for a building mounted small wind turbine. For the evaluation of the impact of the buildings on small wind turbines and human beings it was important to choose a garage made of concrete to observe similar propagation characteristics for structure borne noise and oscillations as residential buildings. The gross floor area of these garages measures 6 m in length and 3 m in width whereas the turbines are installed in the centre of the building at 7.5 m hub height in each case. Figure 4 pictures the two roof types in side view including the masts and spot heights. The gabled roof reduces the distance between the roof and the small wind turbine for about 1.5 m.

As shown in the sketch in Fig. 5, the mast on the building with gabled roof is not connected to the building itself but to the roof framework. This is done to assess

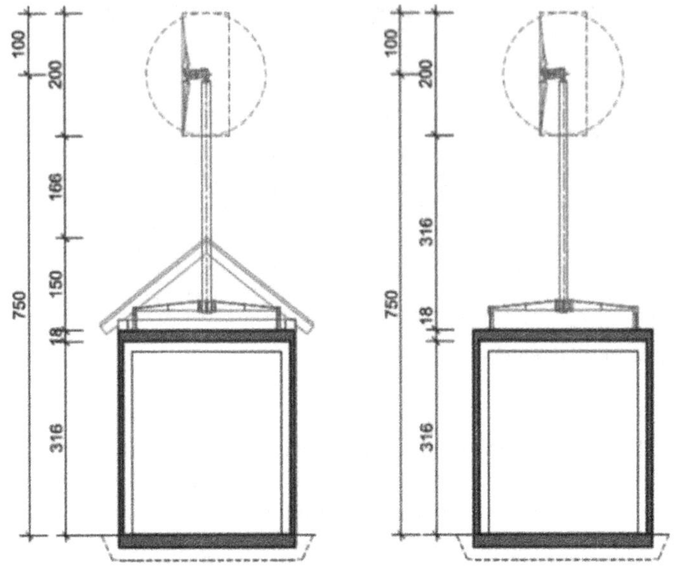

Fig. 4 Building with gabled and flat roof (dimensions in cm)

the damping effect of the wooden structure on oscillation and structure borne noise. The damping effect is quantified by comparing the oscillations in the building with the mast being connected to the roof structure, to the oscillations measured with the mast being connected to the building with an elastomer absorber based connection and a rigid connection respectively. By analysing the roof structure for structural damage before interchanging the turbines between the various positions every three months, it can be estimated whether this type of mast junction is a feasible and safe method for building mounted SWT. Further reductions of oscillations are achieved and evaluated by the use of an oscillation decoupling element as shown in Fig. 5.

First computational fluid dynamics simulations (CFD) served as base when the measurement points along the measurement axis and fixing their relative distances. Figure 6 illustrates the wind speed distribution with reference to the average annual wind speed of 5 m/s measured in 16.5 m above the ground. On the left side the building with gabled roof and on the right side the building with the flat roof is imaged. The colour scale shows changes in wind speed which propagate far behind the building. These changes also occur up to several meters over the building. For this reason, one measurement point is aligned in front of the building in reference to the main wind direction to measure all inlet parameter in three different heights. To measure the influence of the building, two measurement points are aligned behind the building to evaluate the influence on the building in different heights.

In Fig. 7 the test site is illustrated as a 3D model showing the third stage of the project when the buildings and turbines are installed. To evaluate the influence of the building, the terrain and the roof shape, the project is divided in three stages:

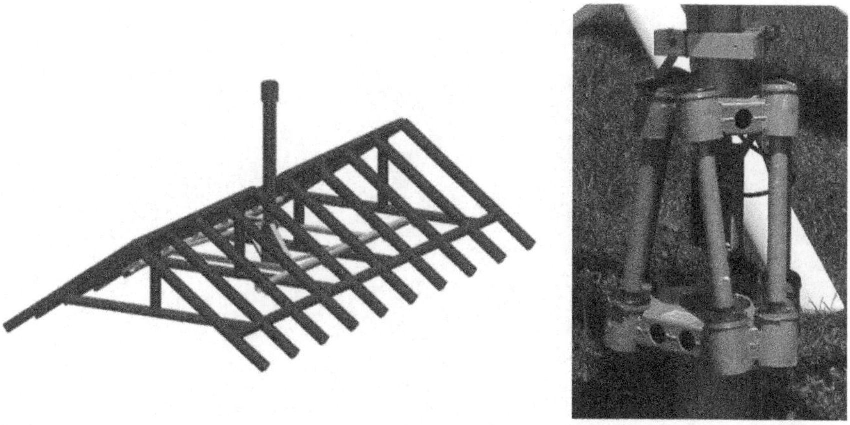

Fig. 5 Mast junction with the gabled roof (left) and oscillation decoupler for VASWT (right)

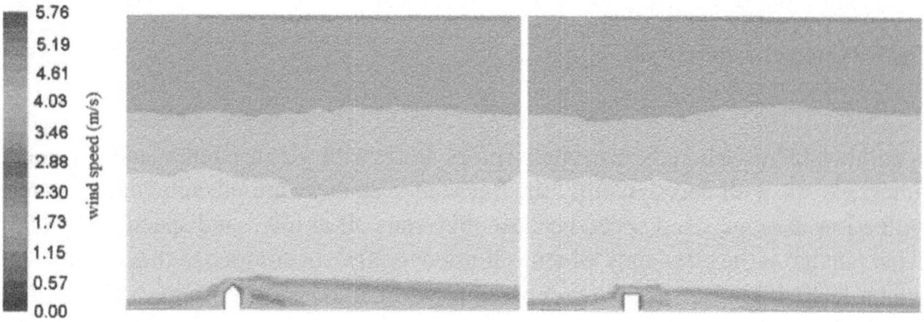

Fig. 6 CFD simulation of the gabled and flat roof

- Site calibration—no buildings are installed but just the measurement masts along the axis.
- Flat roof evaluation—the buildings are installed with the flat roof. The flow pattern is measured and compared to the open field flow pattern.
- Gabled roof evaluation—the building in line with the measurement axis is enhanced with a gabled roof structure and the flow pattern is remeasured.
- SWT test—the SWT are installed and tested on the buildings and the free standing mast.

4 Small Wind Turbines

There are different ways to convert wind energy into kinetic energy in terms of shape and orientation of the rotation axis of wind turbines. The key differentiator is the orientation of the revolution axis. Beside the revolution axis the conversion principle

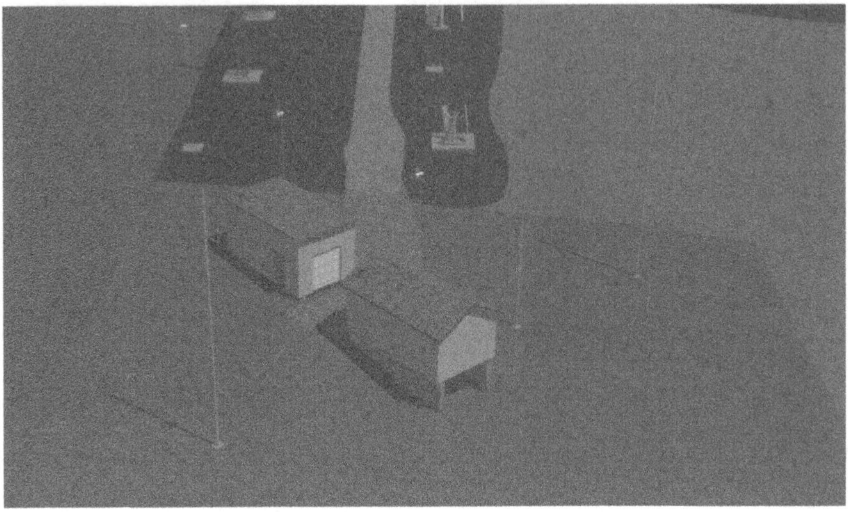

Fig. 7 3D model of the test site

is another factor whereas resistance rotors work with air resistance and buoyancy runners work with aerodynamic lift. Resistance runners are advantageous in areas with a low average wind speed because they start off at low wind speeds. The main disadvantage is the efficiency which is limited to 20%. In contrast to this, a buoyancy runner has a higher efficiency compared to vertical axis SWT, ranging between 35 and 45%. Generally vertical axis small wind turbines have a lower efficiency that horizontal axis small wind turbines but are better suited for use on turbulent sites or sites with frequently altering wind directions [13].

In order to consider the influence of a building on different types of SWTs, the three most common types are deployed and tested in the present investigations. Table 1 shows a summary of all three turbines investigated.

5 Measurement Setup

5.1 Sensor Alignment

The meteorological investigation shall consider the environmental impact in terms of oscillation and vibration as well as noise emissions. Hence, different sensors are deployed to measure all quantities. Figure 8 shows the alignment of different sensors for both cases, gabled and flat roof installation. For the measurement of oscillation there are three different sensor types deployed. High frequency accelerometers marked in red and ranging up to 10 kHz are used to capture high frequency oscillations causing long term fatigue in the structure and being audible for human beings.

Table 1 Investigated small wind turbines

Turbina TE20	Rotor diameter	2 m	
	Weight	280 kg	
	Cut in speed	1.5 m/s	
	Rated speed	18 m/s	
	Type	Vertical axis resistance runner	
	Power	1.2 kW	
Vertkon M	Rotor diameter	2.4 m	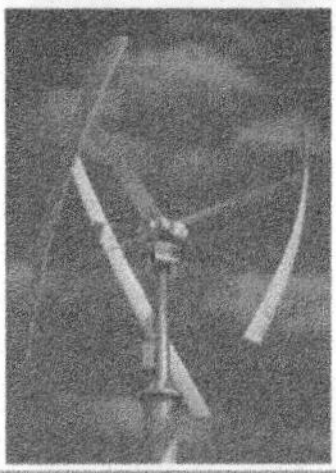
	Weight	150 kg	
	Cut in speed	4 m/s	
	Rated speed	15 m/s	
	Type	Vertical axis buoyancy runner	
	Power	0.95 kW	
FU SkyWind	Rotor diameter	1.5 m	
	Weight	16 kg	
	Cut in speed	4 m/s	
	Rated speed	15 m/s	
	Type	Horizontal axis buoyancy runner	
	Power	1 kW	

Oscillation frequencies below 100 Hz which provoke high loads and stress in the structure are measured with low frequency accelerometers marked in green. Capturing high and low frequency oscillations by separate sensors is necessary as tuning the sampling and analysis algorithms of the used acquisition devices need to be compromised between acquisition speed, resolution and bandwidth. Separate measurement channels allow for better compromising the settings.

Beside the stress for the supporting structure caused by the oscillations of the SWT, the potential influence on human beings living in building is evaluated by recording the oscillation levels in the building with sensors mounted on the floor (marked in blue). Vibrating structures may cause the air to vibrate as well and lead to noise and infra-sound emissions in the building. These emissions are measured with a microphone in the building marked in yellow. Beside these values the rotational speed and the power of each plant is measured.

Oscillation and vibration is first measured in direct proximity to the generator in X and Y dimension with high and low frequency sensors. Previous measurements have shown that oscillation and vibration in the vertical Z dimension are minor due to the high rigidity of the structure in Z axis and can therefore be neglected. As decoupling elements are installed between the generator and the mast along the measurement campaign (see right illustration in Fig. 8), there is also a high and

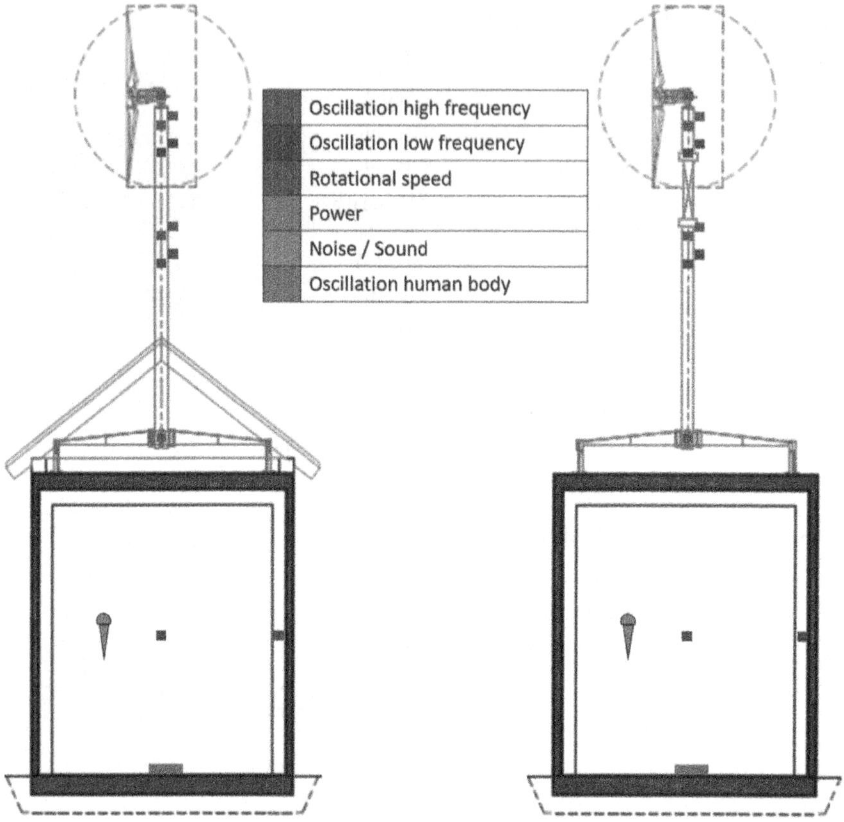

Fig. 8 Sensor alignment

low frequency sensor set installed on the mast to record the reduction potential of these elements. The accelerometer sited at the mast attachment records the incoming vibration before being induced to the building. Comparing this value with the values recorded in the building allows evaluating the damping effect of the junction between mast and building. The junction to the wooden roof structure is expected to give the best results in this case. Seismic oscillations sensors installed on the wall and near the corner of the building measure oscillation and structure borne noise in the concrete mass. Oscillations that propagate from the floor via the feet to the human body are measured with a special standardized sensor and are compared to the maximum tolerated values defined in the EU-directive 2002/44/EG.

5.2 Modal Analysis

Before the measurement of oscillation and vibration of the SWT, the tower and the building a modal analysis is performed to identify all natural frequencies of the system. By stimulating the oscillating system (SWT, mast and building) with a modal hammer and recording the system response spectrum, the towers resonance forms are identified. Knowing the natural frequencies of the system, resonant oscillation events can be recognized and assigned more easily.

5.3 Triggering and Binning

The measurement process is divided in two stages, the first one is a measurement stage only analysing statistical data. Alternated oscillation and vibration is measured with a low sample rate and critical events are analysed. In a second measurement all critical oscillation and vibration intensities are set as trigger events. Then only critical events are recorded with a high sample rate and stored in the belonging bin. When a certain amount of samples for each event is accumulated in a bin the measurement is completed. By this method the amount of recorded data and the expenditure for data processing is limited.

6 Conclusion

When planning and erecting a building mounted small wind turbine, many aspects like the deployed SWT type, building height and shape, oscillations modes and resonances, inflow conditions and others have an influence on the efficiency, safety of operation and service life of a turbine as well as on the quality of life of nearby inhabitants. In the past, many guidelines giving recommendation in these regards, have been published [11, 14, 15].

By validating common CFD-simulation models with real life field measurements, the infrastructure presented in this work provides an opportunity to gain practical experiences and insights to many of the current insecurities in industry and science.

First results have indicated that in fact, the flow velocity and turbulence intensity is influenced by a building and its roof shape. Following investigations and measurements operating the presented setup will give additional comprehension for building mounted small wind turbines and their behaviour in terms of efficiency, oscillation and vibration intensity, acoustic emissions, structure borne sound and fed in power quality. In combination with measurement validated simulation models of the site, detailed understanding for the impact of unsteady inflow conditions on different SWT types can be gained.

References

1. Ragheb, M.: Wind turbines in the urban environment. University of Illinois. http://www.ragheb.co/NPRE%20475%20Wind%20Power%20Systems/Wind%20Turbines%20in%20the%20Urban%20Environment.pdf (2014). Accessed 23 Aug 2017
2. Li, Q.S., Shu, Z.R., Chen, F.B.: Performance assessment of tall building-integrated wind turbines for power generation. Appl. Energy **165**, 777–788 (2015)
3. Ozmen, Y., Baydar, E., van Beeck, J.P.A.J.: Wind flow over the low-rise building models with gabled roofs having different pitch angles. Build. Environ. **95**, 63–74 (2015)
4. Li, Q., Maeda, T., Kamada, Y., et al.: Measurement of the flow field around straight-bladed vertical axis wind turbine. J. Wind Eng. Ind. Aerodyn. **151**, 70–78 (2015)
5. Chong, W.T., Fazlizan, A., Poh, S.C., et al.: The design, simulation and testing of an urban vertical axis wind turbine with the omni-direction-guide-vane. Appl. Energy **112**, 601–609 (2015)
6. Li, Q., Maeda, T., Kamada, Y., et al.: Study on power performance for straight-bladed vertical axis wind turbine by field and wind tunnel test. Renew. Energy **90**, 291–300 (2015)
7. Sanghyeon, K., Cheolung, C.: Development of low-noise drag-type vertical wind turbines. Renew. Energy **79**, 199–208 (2014)
8. Rajasri, K., Bharathi, A., Rajavel, K.: Management of mechanical vibration and temperature in small wind turbines using zigbee wireless network. Int. J. Innov. Res. Comput. Commun. Eng. **7**(2), 5003–5008 (2014)
9. Mollasalehi, E., Sun, Q., Wood, D.: Contribution of small wind turbine structural vibration to noise emission. Energies **6**, 3669–3691. https://doi.org/10.3390/en6083669 (2013)
10. Mohamed, M.H.: Aero-acoustics noise evaluation of H-rotor Darrieus wind turbines. Energy Lab **65**, 596–604 (2013)
11. Olsen, T., Preus, R.: Small wind site assessment guidelines. In: Technical Report NREL/TP-5000-63696. https://www.nrel.gov/docs/fy15osti/63696.pdf (2015). Accessed 22 Aug 2017
12. Toja-Silva, F., Lopez-Garcia, O., Peralta, C., et al.: An empirical–heuristic optimization of the building-roof geometry for urban wind energy exploitation on high-rise buildings. Appl. Energy **164**, 769–794 (2015)
13. Bel Mabrouk, I., El Hami, A., Walha, L., et al.: Dynamic vibrations in wind energy systems: application to vertical axis wind turbine. Mech. Syst. Signal Process. **85**, 396–414 (2016)
14. Tummala, A., Velamati, R.V., Sinha, D.K., et al.: A review on small scale wind turbines. Renew. Sustain. Energy Rev. **56**, 1351–1371 (2014)
15. Jüttemann, P.: Kleinwindkraftanlagen – Der umfassende Ratgeber zur Windanlage für das Haus (1). CreateSpace Independent Publishing Platform (2015)

Environmental Influences on SWT Vibrations and Oscillations

M. Peppoloni, K. Leonhartsberger and A. Hirschl

Abstract Falling energy prices and attractive governmental funding as well as the desire for energy autonomy of private households and enterprises are promoting the expansion of renewable energy generation units. Alongside photovoltaic systems, small wind turbines (SWT) represent one of the few feasible ways to generate electricity in areas with a high building density and in urban areas. In this way, electrical energy can be generated on or near buildings. In comparison to photovoltaic systems, small wind turbines face additional challenges such as the influence on the immediate area due to rotating components. While some environmental impacts are taken into account during the approval procedure, others such as vibrations and oscillations are not considered. A roof-top installation of a SWT is mainly the only feasible way to integrate wind turbines in urban areas with a high building density and guarantee a laminar flow [1]. During the operation of SWTs, many factors such as mass imbalances, aerodynamic imbalances (pitch tolerance), inclination of wind, turbulences and gusts may result in inhomogeneous torque or load, leading to vibration and oscillation of the entire system. Although all types of wind turbines are subject to oscillation, the oscillation of building-mounted SWTs must be seen critically because they can be transduced to the building via the mast. In order to take this environmental impact into account, the "Urbane Windenergie" research project (FFG-number 845184) deals among other issues with the vibration and oscillation of horizontal axis and vertical axis (Darrieus-rotor) small wind turbines. By means of accelerometers, oscillations were metrologically recorded in direct proximity to the generators of two different plants in the Lichtenegg energy research park in Lower Austria. In the first measurement campaign, oscillations were recorded on a vertical axis small wind turbine with a decoupling element installed between the generator and the mast to reduce oscillation intensity. The second measurement, on a horizontal axis turbine without a decoupling element, served to illustrate the different oscillation behaviour of these two turbine types. The evaluated measurement values indicate that the type

M. Peppoloni · K. Leonhartsberger · A. Hirschl (✉)
Department of Renewable Energy, University of Applied Sciences Technikum Wien, Vienna 1210, Austria
e-mail: hirschl@technikum-wien.at

L. Battisti and M. Ricci (eds.), *Wind Energy Exploitation in Urban Environment*,
Green Energy and Technology, https://doi.org/10.1007/978-3-319-74944-0_5

of small wind turbine influences the emitted oscillations and that these oscillations must be considered in particular for turbines for roof assembly.

Keywords Small wind turbine · Vibration and oscillation
Decoupling element

1 Introduction

The fall in energy prices paired with attractive governmental funding and the desire for energy autonomy of private households and enterprises, together with the EU climate targets, are promoting the expansion of renewable energy generation. The EU building directive (Energy Performance of Building Directive, EPBD) in particular encourages a development that makes renewable power plants into a mass application.

Alongside photovoltaic systems, small wind turbines represent one of the few feasible ways to generate electricity in areas with a high building density and in urban areas. In this way, electrical energy can be generated on or near buildings, thus contributing to the EPBD targets. In combination with the desire for energy autonomy, small wind turbines are becoming attractive for private households and are being installed in areas with a high building density.

As compared with PV, small wind turbines face additional challenges such as the influence on the immediate area due to rotating components. Environmental impacts like noise emission, shadow flicker and safety requirements for instance have to be considered so as to reduce the impact on the quality of life. This consideration becomes more important as small wind turbines are installed close to and in inhabited areas. While some environmental impacts are taken into account during the approval procedure, others such as vibrations and oscillations are not considered due to a lack of awareness and information. Not only may roof-top installed small wind turbines in particular involve health issues and influence the quality of life, oscillations transmitted from the mast to the building may also cause long term fatigue of the building structure.

In order to take this environmental impact into account, the "Urbane Windenergie" research project (FFG-number 845184) deals among other issues with the vibration and oscillation of horizontal axis and vertical axis (Darrieus-rotor) small wind turbines. In the absence of normative information about the implementation of oscillation measurements on small wind turbines, an independent measurement concept was drawn up. A part of this concept is based on experiences with oscillation detection concepts for large-scale wind turbines. By means of accelerometers, oscillations were metrologically recorded in direct proximity to the generators of two different plants in the Lichtenegg energy research park in Lower Austria. In the first measurement campaign, oscillations were recorded on a vertical axis small wind turbine with a decoupling element installed between the generator and the mast to reduce oscillation intensity. The second measurement, on a horizontal axis turbine without a decoupling element, served to illustrate the different oscillation behaviour of these two turbine types. The evaluated measurement values indicate that the type of small wind turbine influences the emitted oscillations and that these oscillations must be considered in particular for turbines for roof assembly.

1.1 Problem

Due to small wind turbines (SWT) being smaller than conventional wind turbines, use can be made of the energy potential of wind in built-up areas. A roof-top installation of a SWT is mainly the only feasible way to integrate wind turbines in urban areas with a high building density and guarantee a laminar flow [1].

During the operation of SWTs, many factors such as mass imbalances, aerodynamic imbalances (pitch tolerance), inclination of wind, turbulences and gusts may result in inhomogeneous torque or load, leading to vibration and oscillation of the entire system. Although all types of wind turbines are subject to oscillation, the oscillation of building-mounted SWTs must be seen critically because they can be transduced to the building via the mast. In addition to low frequency oscillations caused by mechanical or aerodynamic asymmetries of the rotor, high frequencies emitted by the generator can be transduced via the mast to the foundation or the building. As a consequence, load-bearing components and the whole supporting structure is stimulated. These components have a range of different natural frequencies that can be stimulated by the turbine and in the end lead to destructive resonant oscillation [1].

The current solution for this issue is the installation of damping elements between the generator and the mast, also known as decoupling elements. Although oscillations on SWTs can have a significant impact on the immediate environment, in many cases the oscillation intensity of a plant is not taken into account and decoupling elements are rarely installed [1].

1.2 Aim of Study

Solutions for vibrating plants are required to ensure a safe and reliable operation of small wind turbines in urban and inhabited areas. The "Urbane Windenergie" (urban wind energy—FFG-number 845184) research project addresses this requirement by assessing and comparing the oscillation intensity of a vertical-axis (VASWT) and a horizontal-axis small wind turbine (HASWT) with similar size and power output. The prior focus of the presented investigation is on wind turbines in urban and inhabited areas. In order to reconstruct a common situation for roof-mounted VASWT and to achieve a safe and reliable operation of the VASWT (Darrieus-rotor), a patented oscillation decoupling element developed and engineered by Dipl.-Ing. Hans Banzhaf was installed. This was not implemented for the HASWT because this type of turbine is usually not fitted with an oscillation-reducing element. Measurements of noise emissions and the evaluation of the economic aspect are not covered in this paper.

In detail, the work described in this paper aims at:

- Evaluating the frequency and oscillation velocity range of the oscillation and vibration occurring.
- Assigning spectral shares of oscillation to their stimulating causes.
- Comparing the oscillation behaviour of a VASWT and a HASWT.

2 Measurement Concept

The evaluation of vibration and oscillation on small wind turbines requires an adequate measurement concept including the alignment of sensors and the choice of bandwidths and the determination of all essential quantities. The measurement of oscillations was conducted by means of piezo-electric and capacitive MEM-sensors measuring the amplitude of the oscillation-acceleration, and inductive sensors measuring the amplitude of the oscillation velocity. These sensors cover a bandwidth of 0.5 Hz–5 kHz. Beside oscillation, various quantities of the wind were determined, partly according to the EN 61400-2 standard. As the averaging period of 1 minute for wind speed defined in the standard is not suitable to analyse brief oscillation events, proprietary averaging had to be defined. The made assumptions are explained in Sect. "2.2. Measured parameters—External influences".

2.1 Small Wind Turbines

2.1.1 Amperius VK250

In the "Urbane Windenergie" research project, an investigation was made of the vibration and oscillation of a VASWT and a HASWT. The selected VASWT was the Amperius VK 250 made by KD Stahl with a nominal power output of 5 kW. The rotor measured 4.5 m in diameter and 5.46 m in height, resulting in a swept area of 24.57 m^2. The plant itself weighed 880 kg and was installed on a tubular mast at a hub height of 20 m (see Fig. 1a).

2.1.2 *Schachner SW05*

The second wind turbine investigated in the project was the Schachner SW05, a downwind horizontal axis turbine (HASWT). The plant was installed on a tubular mast at a hub height of 12 m. Its rotor has a diameter of 5.6 m. The resulting swept

Fig. 1 **a** Amperius VK250 (VASWT), **b** Schachner SW05 (HASWT)

area of 24.63 m^2 and nominal power output of 5 kW are almost equivalent to the VASWT (see Fig. 1b).

2.2 Measured Parameters

An oscillating system is described as the periodic alternation over time of a certain quantity. Parameters such as pressure, temperature, angle and velocity can represent this alternating quantity [2]. Vibrations can be considered as a special type of oscillation as they are periodic alternations which are perceptible by human beings. Hence, vibrations are oscillations of frequencies ranging between 10 Hz and 20 kHz that are induced into the body via hands or feet. When the exposition of a human being to vibration exceeds a certain limit value, health issues may arise [3].

In the present assessment, oscillation acceleration and velocity were selected as the main measurement values, whereat the oscillation frequency was always taken into account. As these parameters allow conclusions to be drawn on the intensity and energetic content of oscillation, they were selected to describe the influence on turbine components.

Different quantities influence the oscillation behaviour of a wind turbine, and so external and internal influences were determined:

- External influences: wind speed, turbulence intensity, inclination of wind.
- Internal influences: rotational speed.

The deflection of a vibrating or oscillating system represented for instance by a spring pendulum can be described with a displacement/time function. The value $\hat{x}$ represents the maximum amplitude of the displacement and ω_0 is the natural frequency of the system. The displacement is calculated as follows:

$$\boldsymbol{oscillation\ displacement}\ \boldsymbol{x(t) = \hat{x} \cdot \sin(\omega_0 t)\ (m)} \tag{1}$$

The differentiation of the function $x(t)$ results in the oscillation velocity/time function, which shows the velocity of a body during one alternation.

$$\boldsymbol{oscillation\ velocity}\ \boldsymbol{\dot{x}(t) = \omega_0 \cdot \hat{x} \cdot \cos(\omega_0 t)\ (mm/s)} \tag{2}$$

Another differentiation of the velocity/time function results in the acceleration/time function, which shows the acceleration of an alternating system from its neutral to its peak position.

$$\boldsymbol{oscillation\ acceleration}\ \boldsymbol{\ddot{x}(t) = -\omega_0^2 \cdot \hat{x} \cdot \sin(\omega_0 t)\ (m/s^2)} \tag{3}$$

Figure 2 shows the qualitative relation between oscillation displacement, oscillation velocity and oscillation acceleration. Depending on the output unit of an oscillation sensor, the desired unit can be calculated by integrating or differentiating. One should note the phase shift, as the oscillation acceleration leads to the oscillation velocity by 90° and to the amplitude by 180° [4].

2.2.1 *Oscillation Acceleration* $\ddot{x}$

The majority of oscillation sensors measure the amplitude of an oscillating system as the acceleration in m/s^2. Triaxial sensors measure oscillations in three orthogonal dimensions. The resulting vectors $\ddot{x}_x$, $\ddot{x}_y$ and $\ddot{x}_z$ rate the three directions respectively. Due to the high rigidity of the mast in the z-orientation, relevant acceleration and velocity values occur mainly in the x and y directions only. For this reason the vertical component (z) was not taken into account in the evaluation. In order to examine the level of oscillation in the light of internal and external influences, root-mean-square (RMS) values $\ddot{x}_{x\_RMS}$ and $\ddot{x}_{y\_RMS}$ were calculated based on a one-second block size with a 75% overlap of each block. The RMS values represent the effective level of oscillation and facilitate a statistical evaluation.

The single axis vectors were then aggregated to a plane oscillation vector $\ddot{x}_{xy\_RMS}$ according to standard ISO 7919-1:1996:

$$\ddot{x}_{xy\_RMS} = \sqrt{\ddot{x}_{x_{RMS}}^2 + \ddot{x}_{y_{RMS}}^2}\ (m/s^2) \tag{4}$$

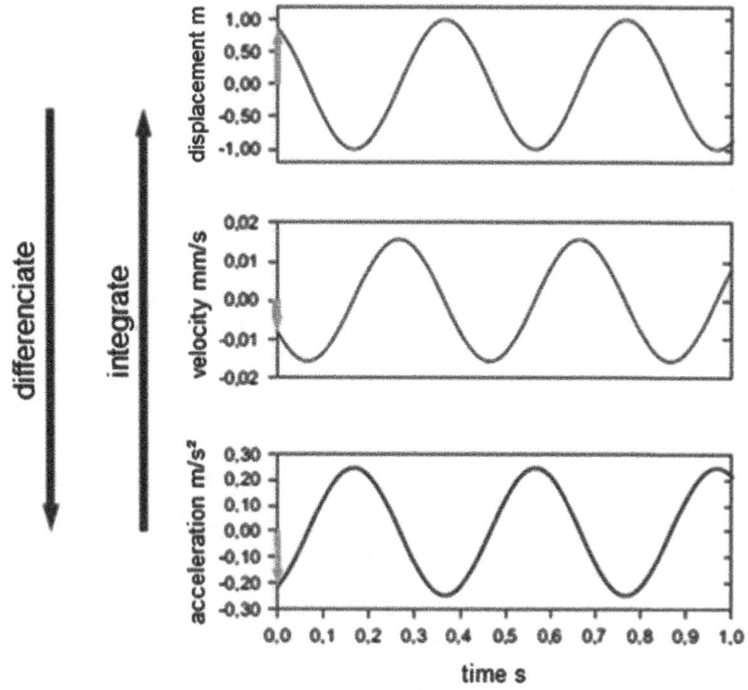

Fig. 2 Displacement, velocity and acceleration [4]

2.2.2 *Oscillation Velocity* $\dot{x}$

Commonly available sensor technology cannot directly measure the oscillation velocity over a wide frequency range. The velocity value in mm/s was obtained by integrating the acceleration signal with a digital data processing function. To obtain a stable output signal, a Butterwoth-type HP-filter with a cut-off frequency of 0.5 Hz had to be used. This filter, however, causes a significant phase shift at frequencies below 5 Hz (see Fig. 3). This effect can be neglected when comparing oscillations of similar frequencies or when adding RMS values as the filter characteristics were identical for all measured velocity channels. With the available data processing functions, a second integration of the signal to obtain the oscillation displacement at frequencies as low as 0.5 Hz would lead to an unstable output signal and was therefore discarded. The resulting RMS-value in plane $\dot{x}_{xy\_RMS}$ was calculated analogous to the oscillation acceleration.

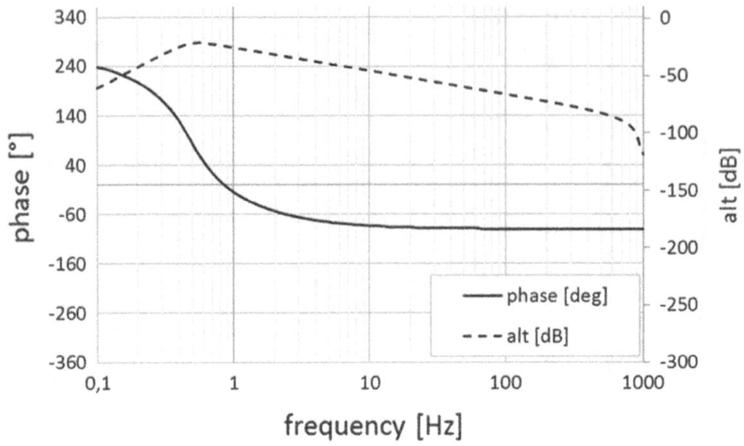

Fig. 3 Frequency response of the integration function including a Butterworth filter 4th order

2.2.3 External Influences

According to EN ISO 61400-2, external environmental influences can be divided into extreme and normal wind conditions. The measurement of these quantities is crucial to show whether the wind has a significant influence on the plant's oscillation behaviour or not. Normal conditions are defined as the wind speed within the operating range of the tested turbine at turbulence intensities up to 50%, while gusts are an extreme external influence. The wind speed was measured by a 3D ultrasonic anemometer in three dimensions v_{w_x}, v_{w_y} and v_{w_z}. The resulting wind signal in plane and in space were calculated with the following formulas:

$$v_{w\_xy} = \sqrt{v_{w\_x}^2 + v_{w\_y}^2}\left(\frac{m}{s}\right) \quad v_{w\_xyz} = \sqrt{\sqrt{v_{w\_x}^2 + v_{w\_y}^2}^{\,2} + v_{w\_z}^2}\left(\frac{m}{s}\right) \tag{5}$$

For the evaluation an average value $v_{w\_AVE\_2}$ in m/s with an averaging time of 2 s (block size) and a block overlap of 75% was calculated.

A turbulent airstream is defined as an abrupt alternation of wind speed taking longitudinal, transversal and vertical alternations into account. The intensity of such a turbulent airstream is calculated by dividing the standard deviation σ in m/s by the average wind speed [5]. As an average value the 20 s average value $v_{w\_AVE\_20}$ was selected instead of the one-minute average value according to the standard. The reason for this was the fact that in a 20 s averaging period certain oscillation events can be assigned more accurately to one turbulent airstream. Accordingly, turbulence intensity was calculated according to the following formula:

$$TI = \frac{\sigma}{v_{w\_AVE\_20}}\,(\%) \tag{6}$$

The vertical angle of the wind direction was calculated relative to three wind dimensions v_{w_x}, v_{w_y} and v_{w_z} with a trigonometric function. In order to determine the inclination, the wind speed on a horizontal plane v_{w_xy} was first calculated with v_{w_x} and v_{w_y}. In a second step of the calculation, the inclination was determined in degrees with the vertical wind speed component v_{w_z}:

$$IN = \frac{arctan\left(\frac{v_z}{v_{xy}}\right) \cdot 180}{\pi} \ (^\circ) \tag{7}$$

According to the IEC-61400-2 standards, gusts are defined as an increase in wind speed of at least 5 m/s within less than 20 s and lasting for more than 3 s [5, 6].

2.3 Sensors and Data Acquisition

For the measurement of oscillation, three different types of sensors were used to cover all frequency ranges. Two sensors measured the oscillation acceleration with capacitive MEMS technology and piezo-electric charge. One sensor directly measured the oscillation velocity via induction. Each sensor showed different characteristics and sensitivities according to the measurement.

The first sensor measured medium to high frequencies between 2 and 5,000 Hz by means of piezo electric charge. The main advantage of this sensor is its accuracy at the previously mentioned frequency range. A disadvantage is the limit for frequencies below 2 Hz, at which significant oscillations can occur. For this reason, a second acceleration sensor with capacitive MEMS technology was used to cover frequencies from 0.5 to 2 Hz. As the high-pass filter, used when integrating the acceleration value into a velocity value leads to damping effects at low frequencies, a geophone was used to record the oscillation velocity directly.

Test measurements showed that only low frequencies ranging between 0.5 and 100 Hz indicate significant amplitudes of the oscillation velocity and that the geophone was not adequate to record oscillations at frequencies below 3 Hz accurately. Accordingly, the capacitive MEM sensor was used for further investigation, a relative measurement error of 20% was accepted for oscillations from 1 to 2 Hz. Oscillations below 1 Hz were not taken into account in the present assessment.

To enable accurate data acquisition, an adequate sample rate for data recording must be selected relative to the maximum expected signal frequency. First tests of the measurement equipment showed that a sample rate 10–20 times higher than the maximum frequency was reasonable and would prevent any aliasing effects. Hence, oscillation acceleration and velocity data was sampled with 2,000 Hz. Signals which alternate more slowly such as wind speed and wind direction were recorded with 200 Hz sample rate in order to reduce the amount of data acquired. Additionally, trigger events for high oscillation, gusts and high turbulence were set at certain limit values:

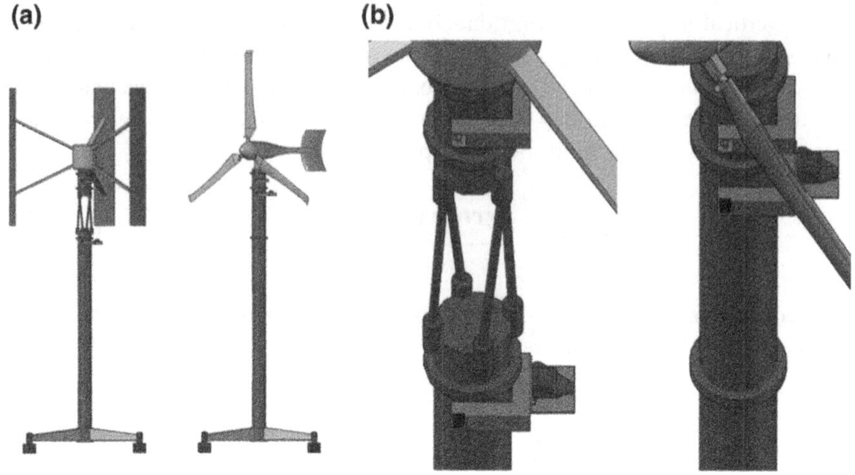

Fig. 4 **a** Sensor alignment on SWTs and **b** measurement layer on SWTs

- Oscillation velocity—trigger value 80 mm/s
- Turbulence—trigger value 25% turbulence intensity
- Gust—wind speed > 5 m/s relative to 20 s average value $v_{w\_AVE\_20}$

When the trigger is not active, oscillation data was stored only in form of RMS and peak values twice a second to reduce the amount of data.

2.4 Sensor Alignment

To capture the maximum oscillation amplitude of the majority of the oscillation modes, MEMS-technology accelerometers were sited at the upper end of both masts. To allow a comparison of the load and stress for the supporting structure, the oscillation intensity considered in this work was captured under the installed damping rod construction for the VASWT. Figure 4a shows a sketch of the VASWT with the decoupling element.

Figure 4b demonstrates how the sensors were connected to the mast. The assembly consisted of an aluminium elbow with a suitable linkage to accommodate the sensors. The assembly was strapped to the mast by a tension belt with a tension greater than 500 N. This setup provided enough rigidity to avoid self-oscillations up to 500 Hz. Oscillations of higher frequency are therefore not taken into account in this work.

Auxiliary piezo-electric and inductive sensors were deployed at nearby locations. First measurements, however, revealed that these sensors were not suitable for capturing the low frequency oscillation generated by the SWT.

The rotor speed was determined by tapping the generator phase and converting the high voltage AC to a 5 V TTL signal. As the frequency of the generator phase

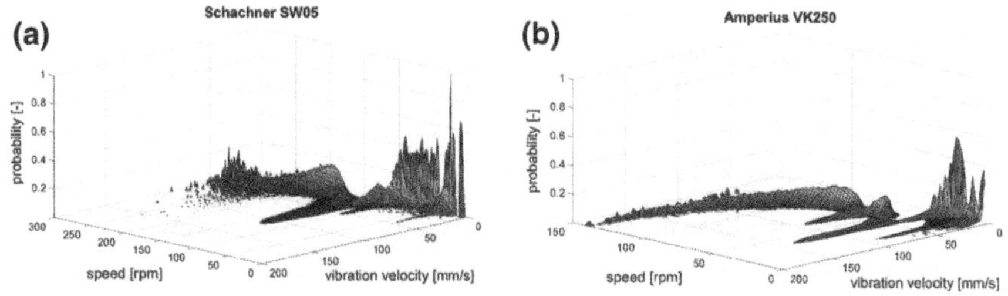

Fig. 5 Rotational speed versus Oscillation velocity **a** HASWT (SW05) and **b** VASWT (VK250)

is proportional to the rotor speed, the rpm can be determined by measuring the frequency of the resulting TTL signal, with the generators pole pair number needing to be factored in.

The wind speed, gusts and the inclination of the wind were measured at a height of 19 m with a 3D ultrasonic anemometer on a preinstalled met mast. The anemometer was located 20 m from the plant.

3 Measurement Results

After prefiltering the datasets of the 16-day measurement operation and excluding downtime periods, the measurements were analysed statistically for an overview of the parameters with the strongest influence on the oscillation intensity of the tested turbines. In a second step the significant sequences with elevated oscillation intensity were analysed in detail and spectral shares of oscillation were assigned to their stimulating cause.

Oscillation velocity was selected as the major evaluation criteria for oscillations below 500 Hz since this parameter is less vulnerable to high frequency disturbance compared to oscillation acceleration and is well distinct at low frequencies.

3.1 Statistical Analysis

In Fig. 5 the oscillation velocity of each oscillation event is plotted in correlation to rotational speed, creating a probability distribution. This statistical evaluation shows the probability of elevated oscillation velocity over the turbine's rpm range. At first sight it can be noted that the oscillation velocity of the (a) HASWT is perceptibly below that of the (b) VASWT.

The diagram (a) for the HASWT (Schachner SW05) shows three peaks marked ranging between 50 and 120 mm/s at a rotational speed of 40, 100 and 130 rpm. These peaks were caused by the stimulation of a resonance or natural frequency

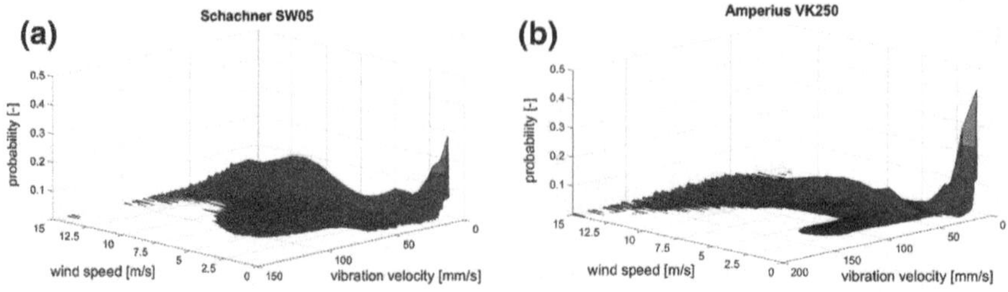

Fig. 6 Wind speed versus Oscillation velocity **a** HASWT (SW05) and **b** VASWT (VK250)

of the mast by the rotor. Near the nominal rotational speed of 240 rpm, oscillation values are scattered and a specific peak value cannot be assigned.

In contrast, the rotational speed of the VASWT (b) (Amperius VK250) caused four peaks ranging between 100 and 200 mm/s over the entire rpm range. Peaks one, two and three are assignable to a specific speed but the fourth increased oscillations appears to build up when approaching the nominal rpm.

These statistical results show a good correlation between rotor rotational speed and increased oscillation intensities. As the rpm is tuned continuously during the operation of an SWT and the frequency of driving forces caused by gyroscopic effects depend linearly on rpm, natural frequencies of certain components are stimulated at dedicated rpm ranges causing resonant oscillations. In Sect 3.2, the rpm ranges at which elevated oscillation velocities were detected are analysed in detail.

As the rotational speed rises with the wind speed, it appears likely that oscillation intensities also correlate with the wind speed. Figure 6 shows the probability distribution of oscillation velocities at specific wind speeds. Figure 6a shows the probability distribution of the horizontal axis wind turbine (Schachner SW05). A peak at 120 mm/s appears at a wind speed ranging between 2.5 and 10 m/s. A comparison of the wind speed and the rotational speed shows that these scattered wind speeds (range 2.5–10 m/s) appear at the critical rotational speed of 130 rpm. This scattered wind peak (see Fig. 6a) represents the oscillation peak at 130 rpm in Fig. 5a, blurred by the fact that 130 rpm can and did occur at different wind speeds. Hence, there is no direct or a weak correlation between the wind speed and high oscillation velocities.

The same effect can be observed in Fig. 6b for the vertical axis wind turbine (VK250). The wind speed is scattered at every peak and these scattered wind speed ranges in Fig. 6b appear at each peak in Fig. 5b. Again, the wind speed indirectly corresponds with oscillation peaks at the vertical axis wind turbine.

A turbulent airstream may appear as an abrupt alternation of the wind speed which is equivalent to a stimulating oscillation. This alternating wind speed can stimulate a natural frequency of a component. For this reason, turbulence intensity was analysed in correlation with oscillation velocity. In addition, the wind speed that corresponds to each turbulence intensity and oscillation velocity was overlaid in colour because the force of a turbulent flow increases with rising wind speed.

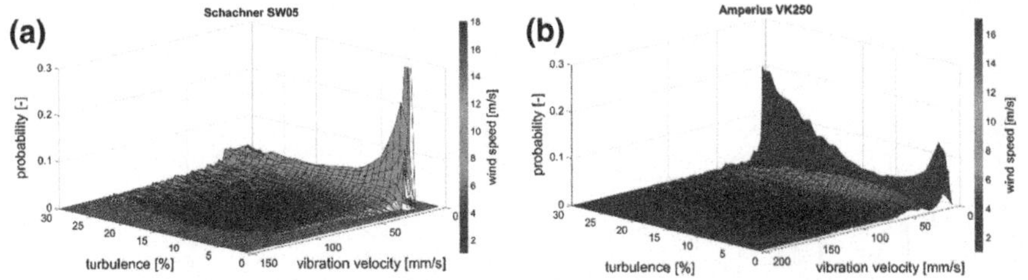

Fig. 7 Turbulence intensity versus Oscillation velocity **a** HASWT (SW05) and **b** VASWT (VK250)

Figure 7 shows the evaluation of turbulence intensity, oscillation velocity and wind speed for the HASWT (Fig. 7a) and the VASWT (Fig. 7b). Figure 7a shows an even distribution of turbulence intensity except for one peak value at a low oscillation velocity ranging between 20 and 40 mm/s. Turbulence intensities between 5 and 30% all have the same distribution of oscillation velocity and wind speed (colour). Figure 7a shows a homogenous probability distribution for all turbulence intensities with velocities around 20 mm/s having the highest probability of around 8%. One exception is a peak at 5% turbulence intensity, 20 mm/s oscillation velocity and a wind speed of 10–12 m/s. As there is few data for low turbulence intensities at high wind speeds, the probability for such an event is relatively high. Thus, this peak corresponds to the scattered rotational speed data at 250–300 rpm in Fig. 5.

Figure 7b shows the influence of turbulences on oscillation of the VASWT showing again an even distribution between 30 and 200 mm/s of turbulence intensity in combination with wind speed. Two peaks at very low oscillation velocities are an exception. These peaks can be considered as an arbitrary distribution of turbulence intensity and oscillation velocity.

Hence, it can be reasoned the turbulence intensity has only minor influence on the level of oscillation velocity.

The inclination of the wind was also analysed for both plants and the results showed a uniform progression of oscillation velocity in correlation to different angles of wind attack. For this reason, the influence of the inclination of the wind was considered as minor like for turbulence.

3.2 *Spectral Analysis*

The statistical analysis shows that rotational speed has the predominant influence on the oscillation intensity of SWT compared to external influences such as wind speed, turbulences and the inclination of the wind. Therefore, the spectral analysis only analyses the correlation between rotational speed and oscillation velocity. As shown in Fig. 5, natural frequencies are stimulated at given rotational speeds causing resonant oscillation of various components of the turbine.

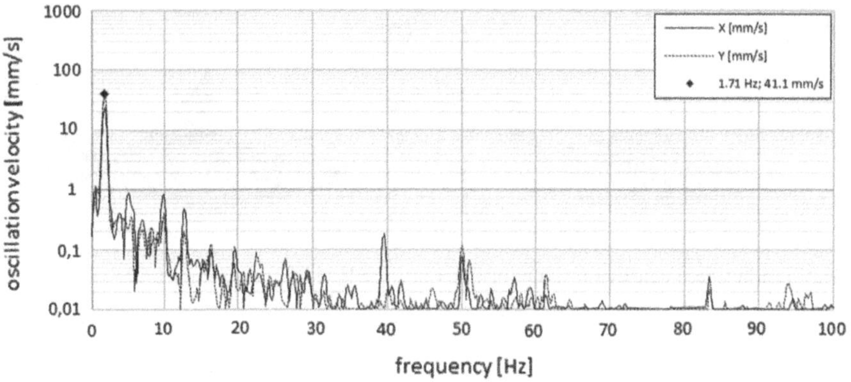

Fig. 8 FFT HASWT at 104 rpm

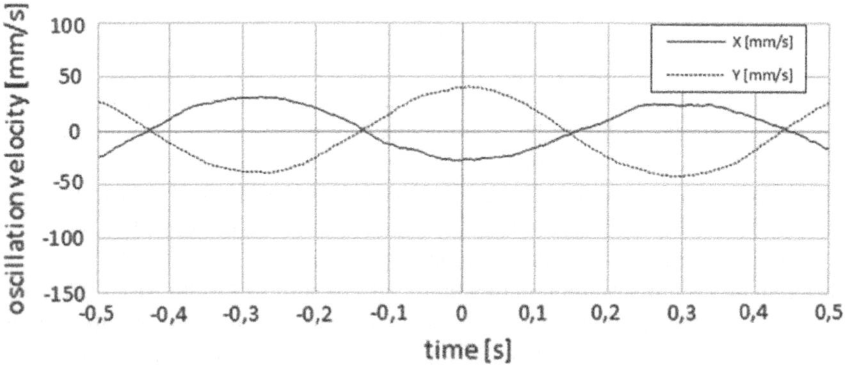

Fig. 9 Scope HASWT at 104 rpm

The first resonant oscillation peak of the HASWT is illustrated in Fig. 8, showing an apparent peak at 1.71 Hz in a Fast-Fourier-Transformation (FFT). The first order oscillation caused by the imbalance of the rotor stimulates a frequency of 1.71 Hz at 104 rpm. The amplitude for the oscillation velocity of the mast is 41 mm/s at 1.71 Hz (1st order). The 3rd and the 16th order caused by the rotor blade passing the tower and the generator poles passing the stator coils respectively, do not emerge from the rest of the signal. This means that these orders do not stimulate any natural frequency of the mast or the plant at this rpm.

The current oscillation velocity over time in Fig. 9 shows the peak value of Fig. 8 at 1.7 Hz. The lines for the X and Y dimensions are phase shifted by 180°, which represents an orbital oscillation of the plant. In addition, the frequency of the oscillation (1.71 Hz) can be observed, as one period lasts for 584 ms.

Figure 10 illustrates the frequency range at 131.5 rpm measured on the mast in direct proximity to the generator. The figure shows peak values in the X and Y planes. The first peak value is at 2.2 Hz at an amplitude of 96.6 mm/s and the second peak

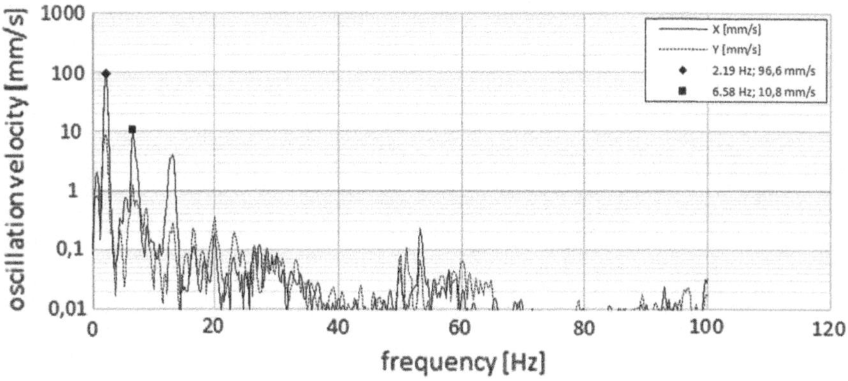

Fig. 10 FFT HASWT at 131 rpm

is at 6.59 Hz at an amplitude of 10.8 mm/s. These peaks represent the first and third order of the rotational speed. For the tested HASWT, there are generally three major sources that cause a periodic displacement of the turbine:

- 1st order—rotor imbalance
- 3rd order—frequency of three rotor blades passing the tower (tower-shadow)
- 16th order—frequency of the 16 generator poles passing the coils (magnetic forces)

At a rotational speed of 131.5 rpm a stimulating frequency (1st order) of 2.2 Hz stimulates a natural frequency of the mast in combination with the turbine. The 3rd order caused by rotor blades passing the tower has a minor impact indicated by the oscillation velocity (10.8 mm/s). At 131.5 rpm, 6.58 Hz is three times the frequency of the rotational speed caused by three rotor blades. The momentum is generated when aerodynamic lift and draft is reduced due to the wind shadow caused by the tower (tower shadow effect). The 16th order does not appear in the FFT. One assumption is that the synchronous generator that has twisted poles along the rotation axis has a smooth operational performance that does not create a significant momentum.

Figure 11 shows an oscillogram of the peak at 131 rpm. The phase shift of 180° between X and Y axis can be observed. Unlike in Fig. 9, the amplitudes of X and Y differ. Additionally, the shape of line X is triangular which indicates an elliptic movement of the tower caused by the first order. The frequency (2.19 Hz) of the first order can be observed, one period of the main oscillation lasting for 450 ms. In addition, the third order is superimposed on the line of the first order with one period lasting for 150 ms which equals 6.58 Hz.

Figure 12 shows one event of the scattered peaks in Fig. 5 of the HASWT between 250 and 300 rpm. In this case the plant was spinning at 270.4 rpm, generating a stimulating frequency of 4.5 Hz. This first order does not appear in first place in the FFT chart. In addition to the first order at 4.5 Hz and the third order at 13.52 Hz there is a peak at 1.7 Hz with an amplitude of 70.9 mm/s which does not correspond to any stimulating frequency occurring in this operation range. Figure 13 shows this

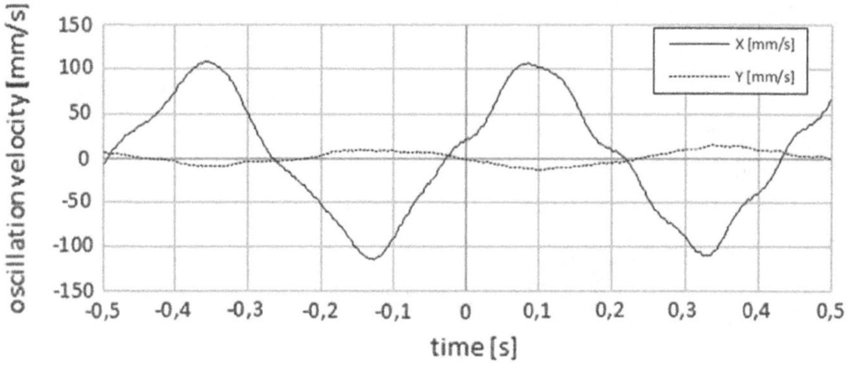

Fig. 11 Scope HASWT at 131 rpm

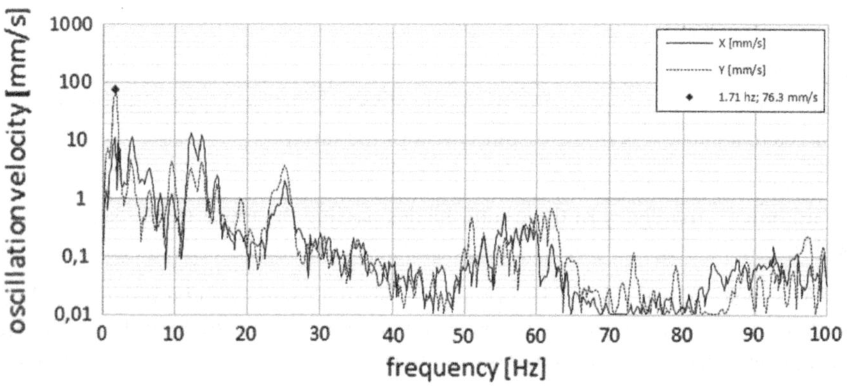

Fig. 12 FFT HASWT at 270 rpm

oscillation at 1.7 Hz, since this event cannot be assigned to any stimulating order, it is likely that this oscillation component is caused by an interference of two oscillations leading to a local resonance in the structure. This is however only an assumption and cannot be verified with the data information.

In comparison to the HASWT, the VASWT shows a similar behaviour at critical rotational speeds. As marked in Fig. 5b there are four different peaks at a rotational speed of 15, 43, 67 rpm and a last peak increasing by square until it reaches its maximum at 137 rpm.

15 rpm is the single rotational speed of the rotor while the plant is starting. This specific rotational speed of the rotor results in an oscillation frequency of 0.73 Hz, which stimulates the mast to oscillate with a maximum amplitude of 81 mm/s as seen in Fig. 14. In this case the third order which is the rotor blade passing frequency does not appear in the FFT.

Figure 15 confirms this behaviour shown in the FFT, the main oscillation has an amplitude of 80 mm/s on average and one period lasts for 1,360 ms, which equals a

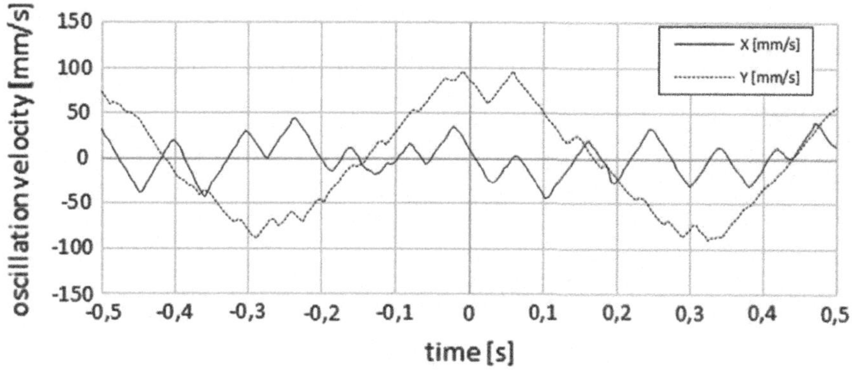

Fig. 13 Scope HASWT at 270 rpm

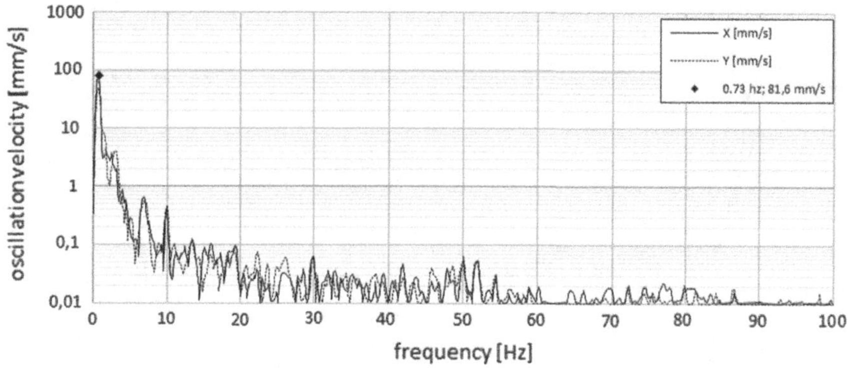

Fig. 14 FFT VASWT at 15 rpm

frequency of 0.73 Hz. The oscillation in the Y dimension is slightly less than the X dimension and phase shifted by 90°, which represents an elliptic movement of the mast.

The second significant peak is shown in Fig. 16 with all three orders marked. At a rotational speed of 43.8 rpm, a stimulating frequency of 0.70 Hz is generated, stimulating the natural frequency of the mast. The third order at 2.19 Hz caused by three rotor blades passing the tower was also stimulated and led to an amplitude of 61.9 mm/s. The 20th order caused by the generator poles does not emerge significantly from the rest of the signal.

The oscillogram in Fig. 17a shows this behaviour at 43.8 rpm on a time scale. The first order is dominant, as indicated by the amplitude and the duration of 1,420 ms of one period. Superimposed on the first order, the third order changes the oscillation significantly, leading to a double-circular movement of the mast as seen in the orbit in Fig. 17b.

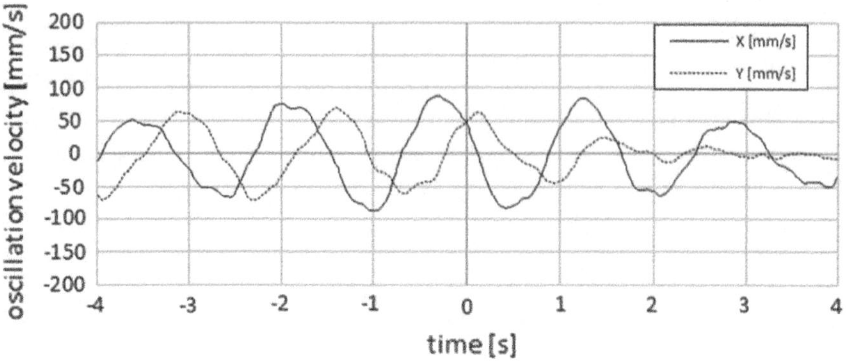

Fig. 15 Scope HASWT at 270 rpm

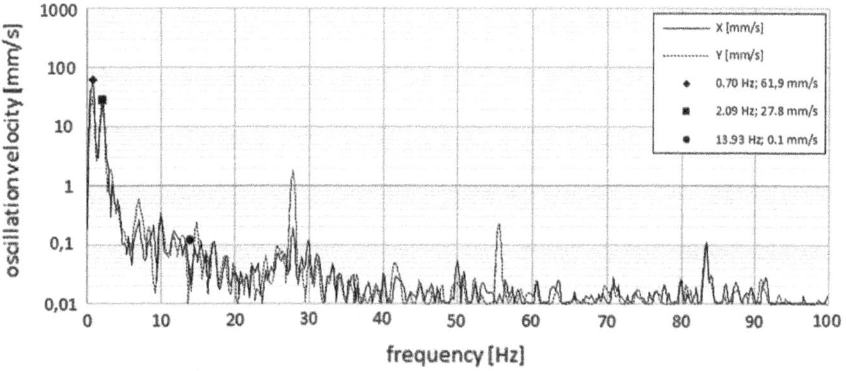

Fig. 16 FFT VASWT at 43.8 rpm

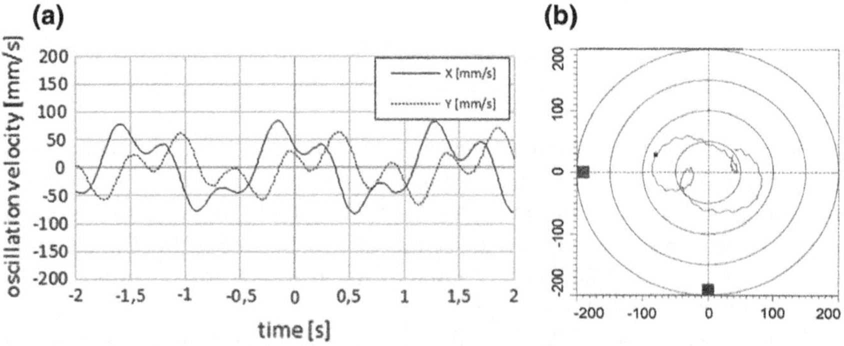

Fig. 17 **a** Scope HASWT at 270 rpm and **b** orbit

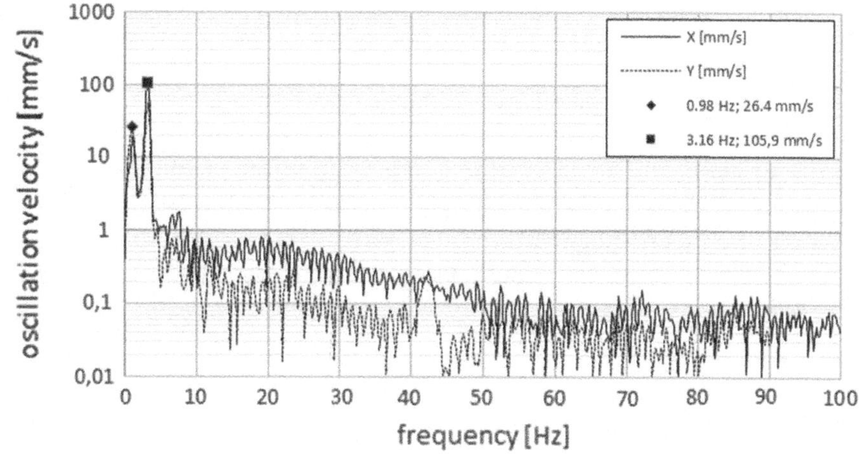

Fig. 18 FFT VASWT at 63 rpm

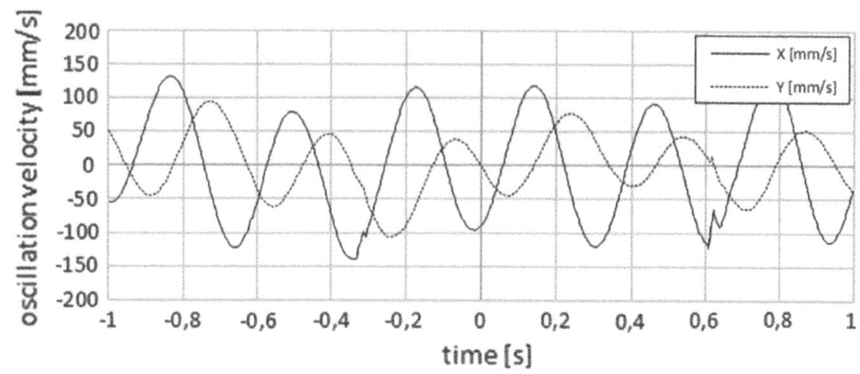

Fig. 19 Scope VASWT at 137.9 rpm

As the rotational speed increases up to 63 rpm, the first order appears at 0.98 Hz with an amplitude of 26 mm/s, but in this case the third order at 3.16 Hz is the most dominant order with an amplitude of 105.9 mm/s, as shown in Fig. 18. In this case the passing frequency of the rotor blades stimulates a natural frequency of 3.16 Hz of a mast component. In this case aerodynamic imbalances of the rotor are the main cause for oscillation.

This predominant third order can also be observed in the oscillogram in Fig. 19. The duration of the dominant oscillation corresponds to 310 ms and an amplitude of 105 mm/s in average to the third order. The first order is superimposed on the third order and observable as a fluctuation of the main oscillation. The Y dimension is lower than the X dimension and phase shifted by 90°, indicating an elliptic movement of the tower.

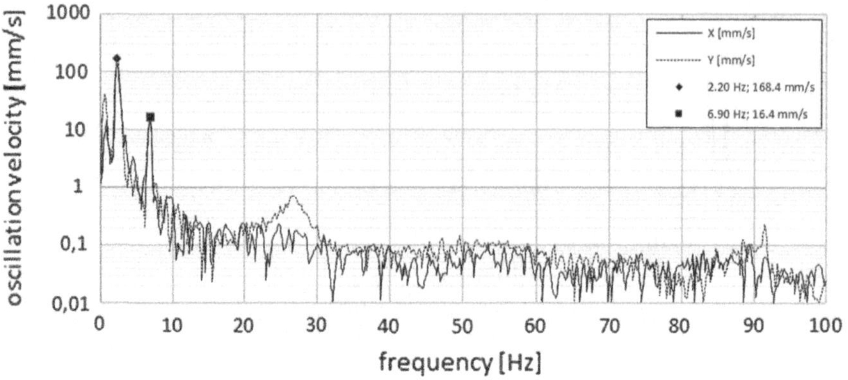

Fig. 20 FFT VASWT at 138 rpm

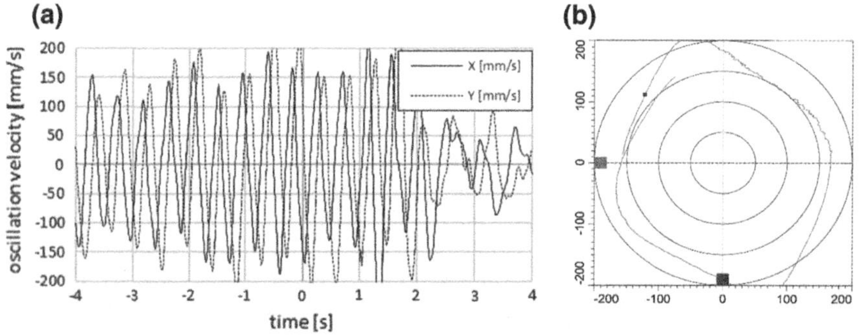

Fig. 21 **a** Scope VASWT at 138 rpm and **b** orbit

Figure 20 shows the last and highest peak of the operational range of the VASWT with a first order of 2.20 Hz, which corresponds to a rotational speed of 138 rpm. Figure 21 again shows the movement of the mast in the scope (a) and orbit (b).

Table 1 shows an overview of all measured oscillation frequencies which could be assigned to a plausible causing effect (order) and showed a significant velocity amplitude, at examined rotor speeds. It will be noticed that the same oscillation frequency of 1.71 Hz, is stimulated by the first order (rotor imbalance) at 100 rpm and at 270 rpm by a non-harmonic effect. For this reason it can be assumed that the frequency of 1.71 Hz is a natural frequency of the mast or other components.

Table 2 shows the same list for the VASWT. In this case two frequencies appear to match at different rotor speeds. The natural frequency of the structure at 0.7 Hz is stimulated at 15 rpm by some non-harmonic effect and at 43 rpm by the imbalance of the rotor (1st order) causing increased oscillations. Another natural frequency of 2.1 Hz is stimulated at 43 rpm by the 3rd order (rotor blades facing upwind) and at 137 rpm by the imbalance of the rotor (1st order).

Table 1 Resonance frequencies HASWT (SW05)

	Rotational speed (rpm)		
	100	131	270
1st order (Hz)	1.71	2.18	4.15
3nd order (Hz)	5.13	6.55	12.21
16th order (Hz)	27.36	34.93	72. 00
Extra order (Hz)	/	/	1.71

Table 2 Resonance frequencies VASWT (VK250)

	Rotational speed (rpm)			
	15	43	63	137
1st order (Hz)	0.3	0.7	1	2.2
3nd order (Hz)	0.8	2.1	3.2	6.9
20th order (Hz)	5	13.9	21	46
Extra order (Hz)	0.7			

4 Conclusion

An analysis of the vibrations and oscillations of a SWT must consider the turbine as a combination with its supporting structure. Since each building has specific natural frequencies which will be stimulated under particular wind flow conditions and rpm ranges by different types of SWTs, an evaluation of a wind turbine (without supporting structure) is of low relevance. This makes the evaluation and interpretation of the vibration and oscillation of building-mounted SWTs complex and difficult to compare. Each supporting structure, be it a mast or a building, has a particular rigidity and range of natural frequencies which will influence the displacement, velocity and acceleration over the oscillation spectrum and thus the behaviour of the turbine. This fact makes it virtually impossible to assess and compare the oscillation of different SWT types installed on different supporting structures.

On the other hand, there are SWTs which cause more oscillations and develop higher force or strain in the structure than others. The two turbines investigated could not be tested on the same tower, an absolute assessment and comparison of the displacement, velocity and acceleration cannot be performed based on the measurement results.

General advice worth following when evaluating the oscillation of a SWT or when choosing a turbine for a specific building is to ensure either overcritical or undercritical operation at higher wind speeds by making sure that the natural frequencies of the supporting structure are not stimulated at an rpm range within 50% or more of the nominal power or wind speed of the turbine. A good tool to do this is a Campbell chart.

An unbalanced mass-distribution of the rotor blades caused by manufacturing tolerances is mostly responsible for the low frequency oscillation of SWTs. However, irregular air flow, turbulences or pitch deviations of the rotor blades result in aerodynamic imbalances. In particular with vertical-axis small wind turbines, aerodynamic imbalances are a systemic problem and should be taken into account. The reason for this is the inhomogeneous rotor blade geometry [1].

References

1. Twele: Empfehlungen zum Einsatz kleiner Windenergieanlagen im urbanen Raum—Ein Leitfaden. HTW Berlin, Berlin (2013)
2. Magnus, K., Popp, K., Sextro, W.: Schwingungen—Eine Einführung in die physikalischen Grundlagen und die theoretische Behandlung von Schwingungsproblemen, 8. Auflage, Vieweg + Teubner Verlag, Wiesbaden (2008)
3. BAUA. http://www.baua.de/de/Themen-von-A-Z/Vibration/Vibration_content.html (2016). Accessed 21 Apr 2016
4. Kuttner, T.: Praxiswissen Schwingungsmesstechnik. Springer Vieweg, Wiesbaden (2015)
5. ÖVE/ÖNORM EN 61400-2:2015-06-01, Windenergieanlagen Teil 2: Anforderungen für kleine Windenergieanlagen (IEC 61400–2:2013)
6. Deutscher Wetterdienst. http://www.dwd.de/DE/service/lexikon/Functions/glossar.html?lv2=100310&lv3=100488 (2016). Accessed 02 Feb 2016

Dynamic Experimental and Numerical Analysis of Loads for a Horizontal Axis Micro Wind Turbine

F. Castellani, M. Becchetti, D. Astolfi and F. Cianetti

Abstract The optimal use of micro wind energy conversion systems is not only a matter of efficiency, but it also involves a number of issues related to the technology and the dimensions of the system in relation with the environment. The growing request of micro wind energy technology for urban areas is nowadays stimulating the research on several subjects. Structural safety is crucial, in order to prevent damages in case of gusts. Noise and vibration assessment and minimization is another relevant issue too, especially for horizontal axis machines with higher efficiency and larger rotational speed. On these grounds, in this work a horizontal axis wind turbine having 2 m of rotor diameter is studied experimentally and numerically. Experimental tests have been performed with steady and unsteady wind conditions and accelerations have been collected, at meaningful operating conditions of the wind turbine, and consequently analyzed. The analysis of the experimental spectra is compared against numerical simulations performed with the aeroelastic code Fatigue, Aerodynamics, Structures and Turbulence (FAST) and this allows to interpret the complex load structure to which the wind turbine is subjected. In particular, evidence of the blade-passing phenomenon is collected: due to the small size of the system, the tower and blades undergo an interesting mechanical interplay.

Keywords Wind energy · Small wind turbine · Aerodynamics · Aeroelasticity Vibrations

F. Castellani (✉) · M. Becchetti · D. Astolfi · F. Cianetti
Department of Engineering, University of Perugia, Via Duranti, 06125 Perugia, Italy
e-mail: francesco.castellani@unipg.it

M. Becchetti
e-mail: matteo.becchetti@unipg.it

D. Astolfi
e-mail: davide.astolfi@unipg.it

F. Cianetti
e-mail: filippo.cianetti@unipg.it

L. Battisti and M. Ricci (eds.), *Wind Energy Exploitation in Urban Environment*, Green Energy and Technology, https://doi.org/10.1007/978-3-319-74944-0_6

1 Introduction

The attitude of public opinion and policy makers towards micro wind energy systems [22] is particularly demanding. Micro wind turbines should be sufficiently simple (and therefore they should have a sufficiently advanced control system) to be handled by citizens without specific expertise. They should convert wind kinetic energy with a certain efficiency, in particular to counteract prejudices [7] on this kind of technology, according to which it performs intrinsically poorly. Their use in urban environment [4–6, 15] poses further challenges: strongly varying loads to which micro turbines, especially in case of gusts [1], are subjected, can cause heavy stresses on all the components, with large noise and vibrations and risks of damage and breakage of the device. Structural safety [19] is therefore a pressing need. Optimizing power extraction means, especially for horizontal axis wind turbines (HAWT), increasing rotational speed and this can introduce vibration and noise problems [9, 14, 16, 21], to which customers and policy makers are particularly susceptible in urban environment. For all these reasons, a deep knowledge of the dynamic behavior of micro wind turbines [10] is fundamental to develop mitigation strategies of noise and vibrations and the present work is a contribution to this topic. This work deals with an HAWT having 2 m of diameter: its design has been optimized through numerical modeling and wind tunnel testing at the University of Perugia [18]. The excellent power extraction and therefore the high rotational speeds reached by the test case HAWT make it the ideal candidate for studying noise and vibration and the possible role of the control strategy for mitigating. This requires a careful awareness of mechanical loads and spectrum [2, 20] and this is actually what is pursued in the present work. An experimental campaign is conducted at the "R. Balli" wind tunnel facility of the University of Perugia.[1] Acceleration measurements are collected at meaningful points of the wind turbine under different working conditions (steady and variable wind flow) and the spectra are interpreted through numerical modeling. Actually, the aeroelastic code FAST [8] is employed to simulate the mechanical response of the test case wind turbine. The structure of the work is therefore the following: Sect. 2 is devoted to the description of the methods and the facilities (numerical and experimental). In Sect. 3, the results are collected. Conclusions are drawn and some further direction is sketched in Sect. 4.

2 The Facilities and the Methods

2.1 The Wind Tunnel

The wind tunnel "R. Balli" of the University of Perugia is in a closed loop configuration (Fig. 1). A 375 kW electric motor, with 11 blades in carbon warped shell,

[1] https://www.windtunnel.unipg.it.

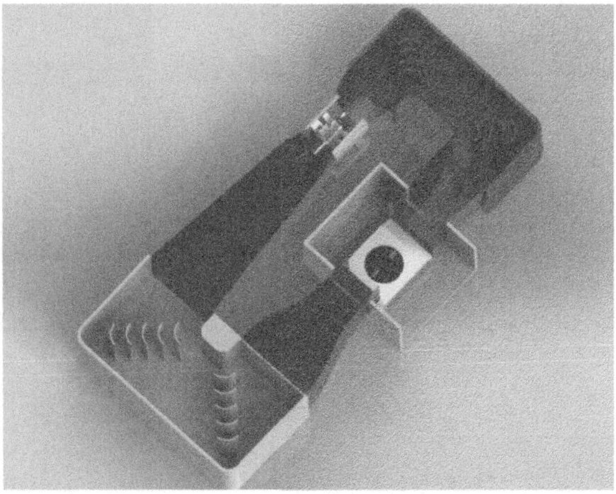

Fig. 1 The wind tunnel facility

accelerates the air up to a maximum speed of 50 m/s. The open test room has approximately 5 m^2 of effective section with a recovery section of about 7 m^2; the turbulence of the flow is conditioned and reduced by means of a honeycomb section. The wind flow is further accelerated by a converging duct into the testing room. Two Pitot tubes measure the wind speed blowing on the test section, and a control station in the testing room measures in real time static air pressure, temperature and relative humidity. The wind tunnel is commonly employed for research activities about small wind turbines [18] and two-wheeled vehicles [11, 12, 17].

2.2 *The Wind Turbine*

The test case wind turbine (Fig. 2) has 2 m of diameter. It is three bladed and the blades are made of glass fiber-reinforced polymer, realized through injection molding. In [18], the design of the blades of this wind turbine has been optimized by an aerodynamic point of view, compatibly with economical constraints (the need of low cost). In [18], it is also shown that the rpm versus power curve of this wind turbine displays significant rpm intervals along which the output is almost flat, supporting the use of the turbine also in considerably turbulent environments. For this study, the wind turbine was controlled using an experimental power curve extracted from previous tests in the wind tunnel. The main reason of this choice was having the wind turbine in its real working condition: this point is also crucial for the FAST simulation, because it is not allowed to simulate the power control of the wind turbine.

Fig. 2 Front view of the test case HAWT in the wind tunnel

2.3 *The FAST Code*

The FAST code has been developed at the National Renewable Energy Laboratory (NREL) of the US government. It is an aeroelastic multibody simulator, developed for modeling two and three bladed HAWTs, offshore and onshore. FAST combines multibody and modal approaches for the dynamical study of the equations of motion defining at each time step the state of the mechanical system. For three bladed HAWTs, as the test case of this work is (Sect. 2.2), 24 degrees of freedom are selected (Fig. 3) for describing the support platform (modeled as a rigid body with 6 dofs), the tower (modeled as Euler-Bernoulli beam), the nacelle (modeled as a rigid body with 1 dof), the transmission and the generator, the furling effect, the blades.

In Fig. 4, the flow chart for the FAST simulation is reported. The primary input file calls all the other files containing information about the blades, the physical and mechanical properties of the tower, and how to model them (Aerodyn file).

2.4 *The Methods*

The HAWT has been instrumented with accelerometers. In particular, as Fig. 5 shows, a three-dimensional accelerometer has been mounted below the nacelle at the top of the tower, and a mono axial accelerometer has been mounted on the tail.

In FAST, it is not possible to simulate the behavior of the tail of the turbine, so a purely experimental vibration analysis was performed in order to understand the criticality of the tail itself. The accelerometer placed on the tail has been employed for collecting measurements under a wind ramp from 6 to 11 m/s, reported in Fig. 6.

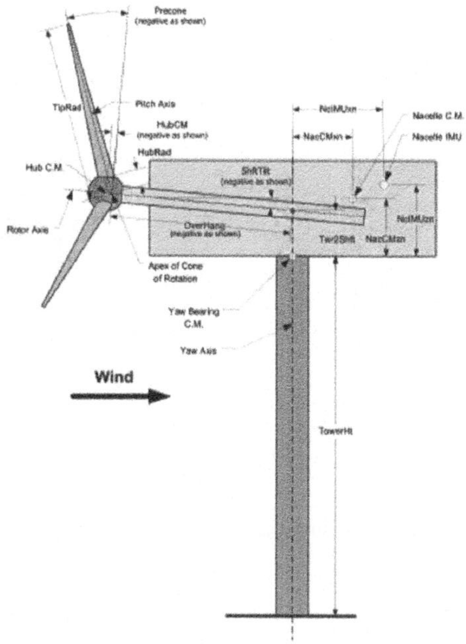

Fig. 3 Configuration of a three-bladed HAWT in FAST

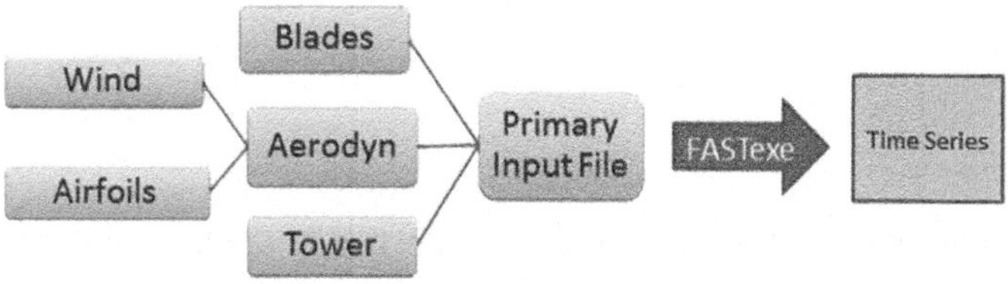

Fig. 4 Flow chart for FAST simulation

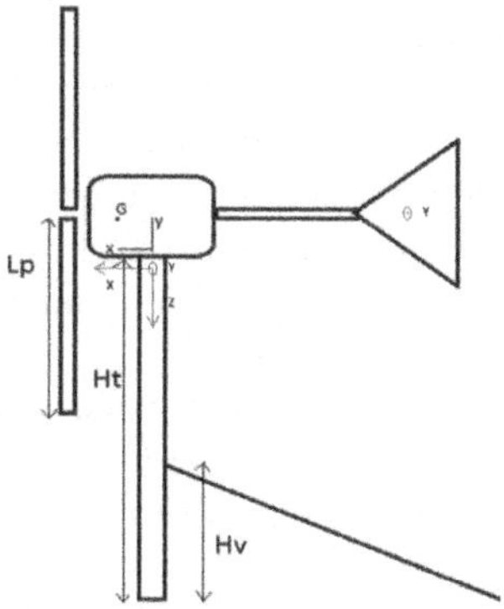

Fig. 5 Reference axes for FAST Tower references (black) and reference axes for accelerometer measurements (red) (color figure online)

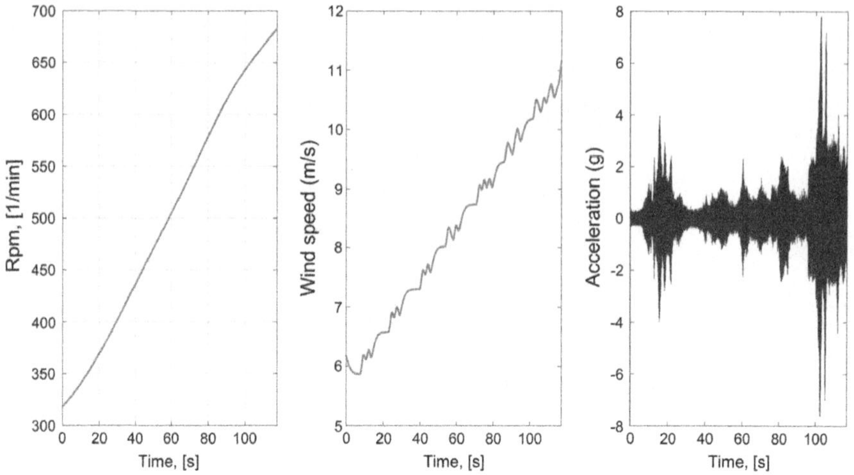

Fig. 6 Revolutions per minute, wind speed and acceleration signal for the analysis of the tail of the wind turbine

Table 1 Dimensions of the system, according to Fig. 5

Quantity	Dimension (m)
Lp	1
Ht	0.98
Hv	0.55
XG	0.13
YG	0.2

Concerning the tower vibrations, the objective is to compare the experimental data against the FAST simulations, especially as regards the frequency content and its interpretation. For doing this, a careful set up must be fine tuned. The structure of the tower and its fixing is shown in Fig. 5. The dimensions of the characteristics parameters of the turbine are shown in Table 1.

In order to accept the input file data for the tower, FAST needs the dimensions and the Rayleigh coefficient to compute the modes of the structure.

The natural frequencies of the tower have therefore been measured using an hammer as excitation, with the turbine stopped. Using ANSYS [13], a model of the tower was reconstructed and tuned using the information for the natural frequencies collected through these experimental tests.

In the following Fig. 7, an example is reported of a wind intensity signal employed for the experimental tests. This kind of signal is reproduced numerically and provided as input to the FAST environment, in order to compare the simulations to the experimental results.

After the first trials, it has been observed that the influence of the tower is very relevant for the matching of numeric ad experimental data. The FAST code has actually been developed for large wind turbines, where the distance from the blade to

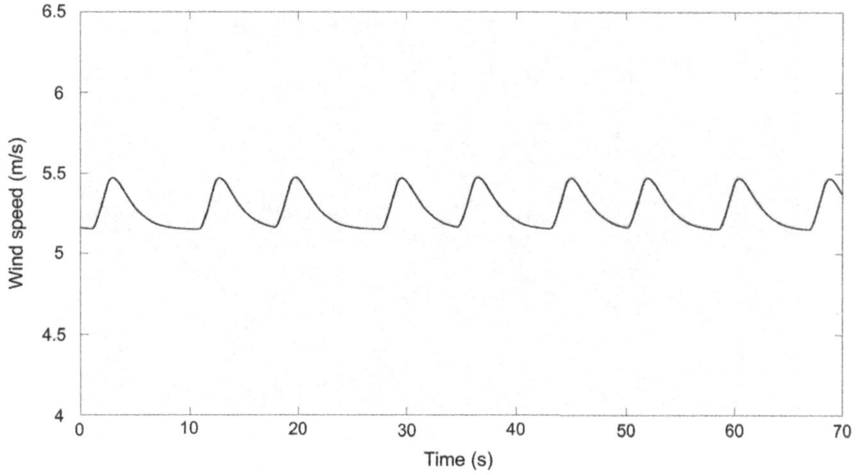

Fig. 7 Experimental wind speed signal

the tower is so large that it is not relevant by a mechanical point of view. For the micro wind turbine under investigation, the blade-tower distance is such that the blade-passing phenomenon [3] is relevant, i.e. the mechanical interplay between tower and blades. This happens especially at high wind intensities, when the flexibility of the blade pushes the distance from the tower to its minimum.

3 The Results

3.1 Tower Analysis

The comparison of numerical against experimental results has been focused on the x oscillations of the tower (Fig. 5), as a reference for auto-tuning of the post-processing. First of all a frequency analysis has been performed and the results are collected in Fig. 8, where the main frequency inputs of the acceleration signals are highlighted.

To properly interpret Fig. 8, it is important to recall what are the main loads transmitted to the tower and their frequency contribution. The main loads are divided into mechanical loads and aerodynamic loads. Mechanical loads can be static loads or dynamic loads: the former are constant over time and cause a constant deflection linked to the rigidity of the structure. Static mechanical loads are therefore essentially due to the force of gravity. Dynamic loads are related to the motion of the rotor and the interplay with the control systems. If the rotor is unbalanced (i.e. it has uneven mass distribution), the imbalance can be modeled as the presence of a mass material point m at a certain distance r from the rotor. The mass m gives rise to a moment (which is variable in time) with respect to the rotational axis and to a centrifugal

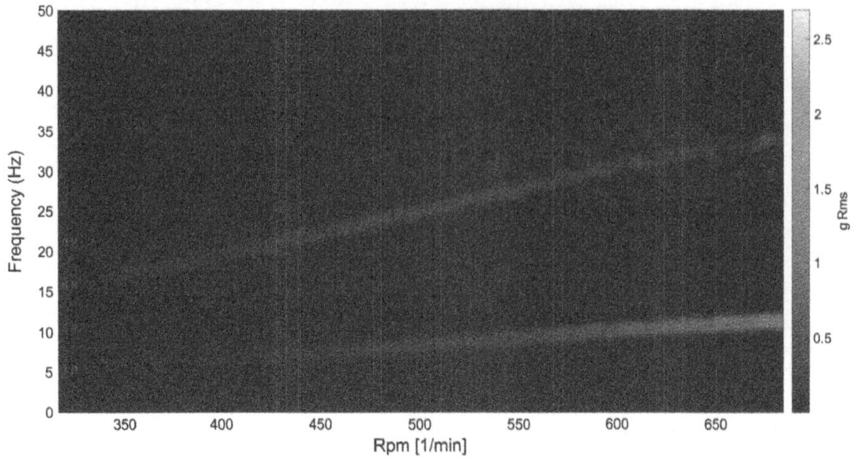

Fig. 8 Frequency analysis on tower acceleration on axes x during ramp test

force having components in the rotor plane and also resulting in bending and torque on the tower. These loads due to imbalance have a frequency component equal to the rotor rpm (1P). Aerodynamic loads can be divided into static loads, cyclic loads and random loads: static loads are caused by constant wind conditions and result in a constant deflection of the structure; random loads arise mainly due to wind turbulence and are one of the major causes of fatigue breaking. Cyclical loads, due to particular aerodynamic effects, are cyclically present in the rotary (blade) reference system and in the fixed reference system of the tower. These loads are crucial for studying the evolution of the Fast Fourier Transform (FFT) of the experimental acceleration measurements. In the rotary reference system (for example at the blade root) the cyclic loads are harmonic (multiples) of the rotor angular velocity. Then there will be loads with inputs of 1P, 2P, 3P and so on. In the fixed reference system (tower base), these loads, due to the presence of the three blades, translate and become multiples of three times the rotor angular velocity: it is therefore possible to see loads with frequency content 0P, 3P, 6P, 9P. In the case of the turbine under investigation, the 1P component is visible in the graph as 6P. The unbalancing of the rotor is linked to 1P. For 6P, the reason is the blade-passing phenomenon. This effect is due to the passage of the shovel near the tower and it is present in both downwind turbines and upwind turbines: it is due to the modified flow conditions at the tower. In the case of downwind turbines, the tower creates a real shadow area and therefore the velocity of the fluid upstream the turbine is consistently diminished when the blade passes near the tower. For these reasons, even if the considered turbine configuration is upwind, the numerical set up has carefully fine tuned as regards damping ratio and tower shadow coefficient in order to include the blade passing interferences induced by the very small blade-tower gap. In Fig. 9, a comparison between FAST simulation and experimental data is provided.

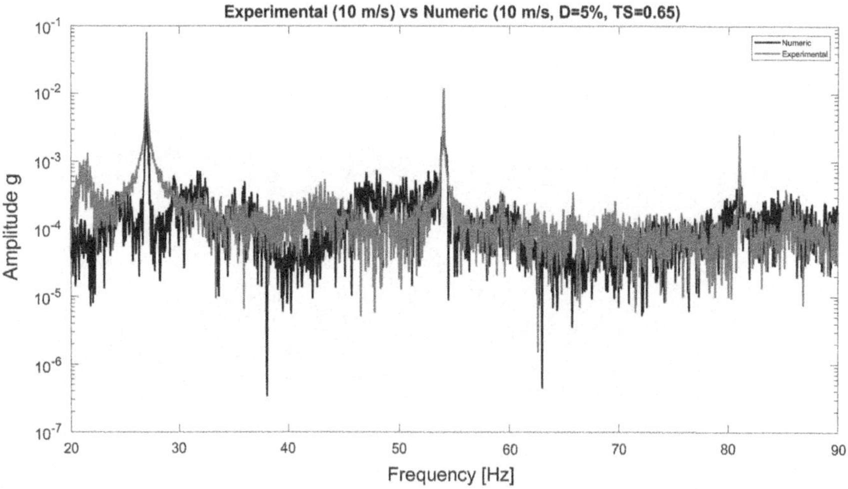

Fig. 9 FFT of numerical and experimental data: 10 m/s of wind speed

Figure 9 refers to the case wind speed equal to 10 m/s and the numerical model is set with $d = 5\%$ and $TS = 0.65$, where d is the damping and TS is the tower shadow coefficient. These coefficient have been tuned using a systematical iteration of FAST solver in order to improve the numerical analysis. Finally numerical and experimental data have a good degree of correspondence in terms of location of aerodynamic peaks 3P, 6P and 9P. The intensity of the peaks is influenced by the tower factor and its fine-tuning deserves further devoted studies.

3.2 *Tail Analysis*

A waterfall (Fig. 10) and order (Fig. 12) analysis has been performed, for investigating the frequency content of the vibration of the tail, linked to the tower vibration. The same analysis has been performed in the y direction of the tower accelerometer (Fig. 11).

As can be seen in Fig. 12, the vibration of the tower is very low and no traces of constant frequencies appear. For the tail waterfall diagram, instead, many frequencies compose the spectrum and remain constant when the rotational speed increases. Comparing Figs. 11 to 10, the interpretation is confirmed, according to which the vibration of the tail induces also vibrations on the tower. In fact, where these lines cross a rotational line (as happens at the 48th order of rotation), the level of acceleration grows up to excite the tail.

This analysis demonstrates that, for these small devices, also secondary elements in term of overall weight (such as the tail), can give a meaningful structural excitations that can induce severe loads in some operational conditions. This can be quickly crosschecked with experimental measurements but can be hardly modeled numerically.

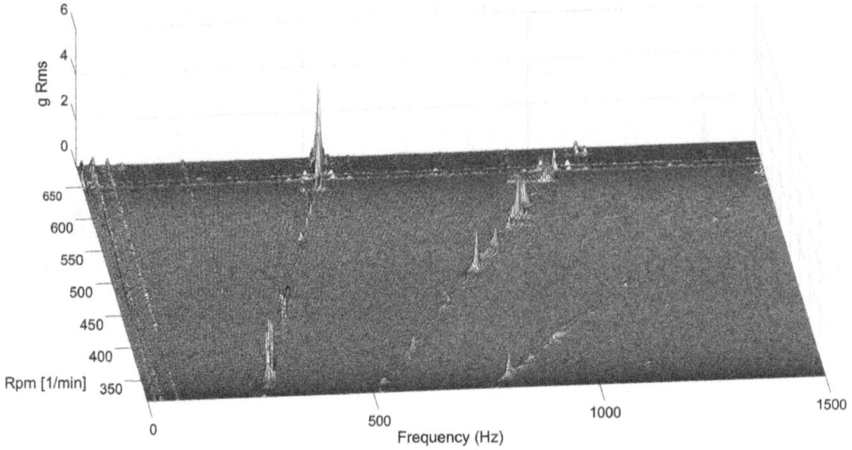

Fig. 10 Waterfall of vibration of the tail in y direction

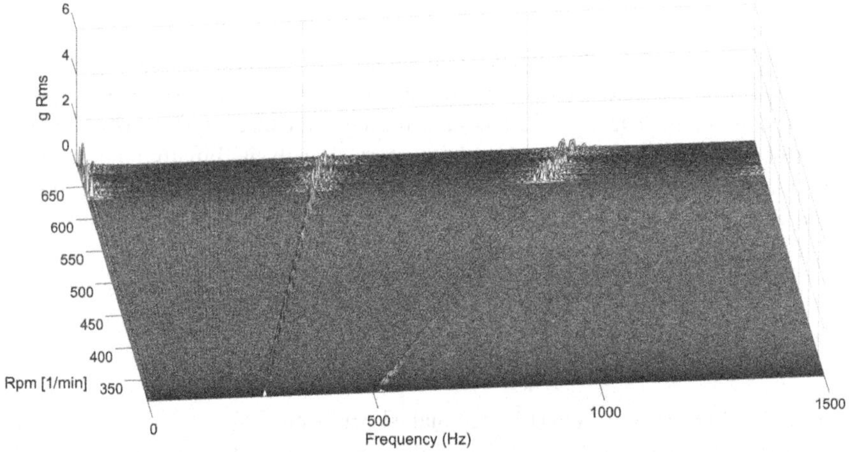

Fig. 11 Waterfall of vibration of the tower in y direction

4 Conclusions and Further Directions

In this work, an experimental and numerical analysis has been conducted on a micro HAWT. The experimental analysis has been carried out at the wind tunnel facility of the University of Perugia and it basically consists in accelerometer-based vibration analysis. On the numerical side, a model of the wind turbine is implemented through the aeroelastic code FAST. The main outcome of this work is that such an integrated approach is not only valuable, but somehow unavoidable, if one wants to construct a numerical framework capable of reproducing and interpreting the complex dynamics between the wind flow and such a small wind turbine. Actually, for example, one of

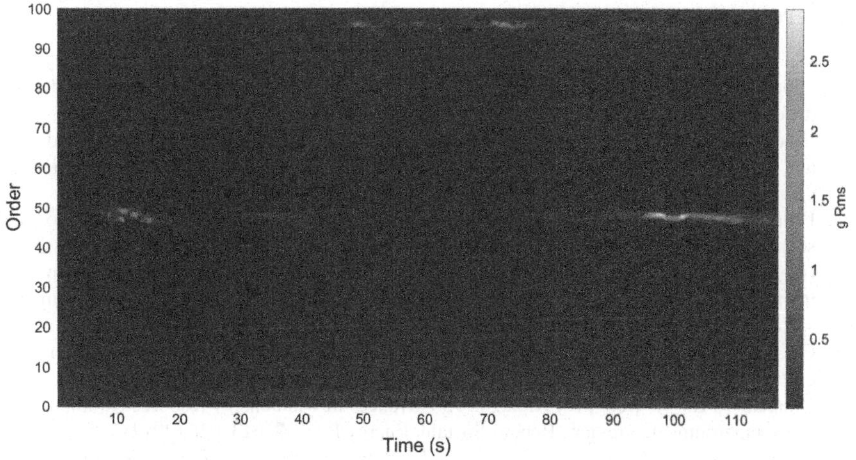

Fig. 12 Order of vibration of the tail in y direction

the main peculiarities of micro wind turbines is the blade-passing phenomenon: it is the mechanical interplay between the blades and the tower and it can be particularly relevant at high rpms, when the flexibility of the blade pushes the relative distance to its minimum and the vibration transmits to the tower. In order to take into account this phenomenon, the numerical model must be fine tuned as regards damping ratio and tower shadowing coefficients and this requires a constant feedback between experiments and computational framework. Actually, the experimental analysis of Sect. 3.2 demonstrates that also secondary elements in term of overall weight (such as the tail) can give meaningful contributions to the spectrum under certain operating conditions but this is very difficult to model, while it is more intelligible to detect and interpret experimentally. On the other hand, once a reliable aeroelastic model of the system has been set up, it becomes particularly useful for interpreting the complex load structure on the system: the frequency content of the measured tower vibrations have actually been discussed comparing to the numerical model (Fig. 9). Sideways, frequencies associated to rotor unbalancing have been observed. Further directions of this work include the assessment of the role of the control system and the interplay with the aeroelastic dynamic structure. Actually, the overall future objective, in the framework of the SOFTWIND research project, is testing and validating control strategies in terms of efficiency in responding to variable speed and capability to relieve dynamic loads. The structural behavior of the system, the stress state of the components and the noise response of the wind turbine shall be addressed for a 360° perspective on numerical modeling and experimental testing for the purposes of micro wind turbine technology.

Acknowledgements This research activity was partially supported by Italian PRIN funding source (Research Projects of National Interest—Progetti di Ricerca di Interesse Nazionale) through a financed project entitled SOFTWIND (Smart Optimized Fault Tolerant WIND turbines).

References

1. Alpman, E.: Aerodynamic performance of small-scale horizontal axis wind turbines under two different extreme wind conditions. J. Thermal Eng. **1**(3), 420–432 (2015)
2. Bradney, D., Evans, S., Da Costa, M.S.P., Clausen, P.: Comparison of computational modelling and field testing of a small wind turbine operating in unsteady flows. In: Journal of Physics: Conference Series, vol. 753, p. 082029. IOP Publishing (2016)
3. Dolan, D.S., Lehn, P.W.: Simulation model of wind turbine 3p torque oscillations due to wind shear and tower shadow. IEEE Trans. Energy Convers. **21**(3), 717–724 (2006)
4. Gagliano, A., Nocera, F., Patania, F., Capizzi, A.: Assessment of micro-wind turbines performance in the urban environments: an aided methodology through geographical information systems. Int. J. Energy Environ. Eng. **4**(1), 43 (2013)
5. Grieser, B., Sunak, Y., Madlener, R.: Economics of small wind turbines in urban settings: an empirical investigation for germany. Renew. Energy **78**, 334–350 (2015)
6. Ishugah, T., Li, Y., Wang, R., Kiplagat, J.: Advances in wind energy resource exploitation in urban environment: a review. Renew. Sustain. Energy Rev. **37**, 613–626 (2014)
7. James, P., Sissons, M., Bradford, J., Myers, L., Bahaj, A., Anwar, A., Green, S.: Implications of the UK field trial of building mounted horizontal axis micro-wind turbines. Energy Policy **38**(10), 6130–6144 (2010)
8. Jonkman, J.M., Buhl Jr., M.L., et al.: Fast users guide. Technical report REL/EL-500-38230, National Renewable Energy Laboratory (2005)
9. Lee, S., Lee, S.: Numerical and experimental study of aerodynamic noise by a small wind turbine. Renew. Energy **65**, 108–112 (2014)
10. Lubitz, W.D.: Impact of ambient turbulence on performance of a small wind turbine. Renew. Energy **61**, 69–73 (2014)
11. Mariani, F., Poggiani, C., Risi, F., Scappaticci, L.: Formula-sae racing car: experimental and numerical analysis of the external aerodynamics. Energy Procedia **81**, 1013–1029 (2015)
12. Mariani, F., Risi, F., Bartolini, N., Castellani, F., Scappaticci, L.: Spoilers optimization to reduce the induced stresses on a racing helmet. Technical report, SAE Technical Paper (2016)
13. Moaveni, S.: Finite Element Analysis Theory and Application with ANSYS, 3rd edn. Pearson Education India (2008)
14. Mollasalehi, E., Sun, Q., Wood, D.: Contribution of small wind turbine structural vibration to noise emission. Energies **6**(8), 3669–3691 (2013)
15. Pagnini, L.C., Burlando, M., Repetto, M.P.: Experimental power curve of small-size wind turbines in turbulent urban environment. Appl. Energy **154**, 112–121 (2015)
16. Rahman, M., Ong, Z.C., Chong, W.T., Julai, S., Khoo, S.Y.: Performance enhancement of wind turbine systems with vibration control: a review. Renew. Sustain. Energy Rev. **51**, 43–54 (2015)
17. Scappaticci, L., Risitano, G., Battistoni, M., Grimaldi, C.: Drag optimization of a sport motorbike. Technical report, SAE Technical Paper (2012)
18. Scappatici, L., Bartolini, N., Castellani, F., Astolfi, D., Garinei, A., Pennicchi, M.: Optimizing the design of horizontal-axis small wind turbines: from the laboratory to market. J. Wind Eng. Ind. Aerodyn. **154**, 58–68 (2016)
19. Tabrizi, A.B., Whale, J., Lyons, T., Urmee, T.: Performance and safety of rooftop wind turbines: use of CFD to gain insight into inflow conditions. Renew. Energy **67**, 242–251 (2014)
20. Tabrizi, A.B., Whale, J., Lyons, T., Urmee, T., Peinke, J.: Modelling the structural loading of a small wind turbine at a highly turbulent site via modifications to the kaimal turbulence spectra. Renew. Energy **105**, 288–300 (2017)
21. Taylor, J., Eastwick, C., Lawrence, C., Wilson, R.: Noise levels and noise perception from small and micro wind turbines. Renew. Energy **55**, 120–127 (2013)
22. Tummala, A., Velamati, R.K., Sinha, D.K., Indraja, V., Krishna, V.H.: A review on small scale wind turbines. Renew. Sustain. Energy Rev. **56**, 1351–1371 (2016)

Rooftop Siting of a Small Wind Turbine Using a Hybrid BEM-CFD Model

F. Balduzzi, A. Bianchini, D. Gentiluomo, G. Ferrara and L. Ferrari

Abstract The benefits of wind turbine rooftop installations are related to the exploitation both of a higher elevation within the atmospheric boundary layer and of possible local accelerated flows originated by the interaction between the wind and the surrounding landscape. The selection of the proper turbine positioning is however pivotal to ensure maximized energy yields. Although the complete solution of the flow field surrounding the rotors would lead to most accurate results, lower-fidelity models with a more affordable computational cost are still about to be preferable for multivariate optimization analyses. In this study, a set of simulations using a hybrid BEM-CFD model were carried out to optimize the siting of a small HAWT in the rooftop of a suburban building. The parametric study on the urban landscape and the turbine positioning showed that the proposed approach hybrid approach provides interesting prospects in view of more energy-efficient urban installations of wind turbines.

Keywords Wind turbine · Rooftop · CFD · Built environment · Siting

List of Symbols and Abbreviations

ABL	Atmospheric Boundary Layer
ADM	Actuator Disk Model
BEM	Blade Element Momentum
CFD	Computational Fluid Dynamics

F. Balduzzi (✉) · A. Bianchini · D. Gentiluomo · G. Ferrara
Department of Industrial Engineering (DIEF), Università degli Studi di Firenze, Via di Santa Marta 3, 50139 Florence, Italy
e-mail: balduzzi@vega.de.unifi.it

L. Ferrari
Department of Energy, Systems, Territory and Construction Engineering (DESTEC), University of Pisa, Largo Lucio Lazzarino, 56122 Pisa, Italy
e-mail: lorenzo.ferrari@unipi.it

L. Battisti and M. Ricci (eds.), *Wind Energy Exploitation in Urban Environment*, Green Energy and Technology, https://doi.org/10.1007/978-3-319-74944-0_7

C_μ	Turbulence model constant
C_p	Power coefficient
C_s	Roughness constant
d	Displacement (m)
D	Distance between UB and IB (m)
h	UB height (m)
H	IB height (m)
$\hat{H}$	Mean buildings height (m)
HAWT	Horizontal Axis Wind Turbine
K_s	Sand-grain roughness (m)
k	Turbulent kinetic energy (m^2/s^2)
IB	Installation Building
L	Buildings width (m)
P	Turbine power (W)
R	Turbine radius (m)
R^2	Coefficient of determination
RANS	Reynolds-Averaged Navier-Stokes
TSR	Tip-Speed Ratio
UB	Upwind Building
$u*$	Friction velocity (m/s)
V	Flow velocity (m/s)
VAWT	Vertical Axis Wind Turbine
VBM	Virtual Blade Model
y_p	Height of the ground cells centroid (m)
z_0	Roughness length (m)

Greek Letters

γ	Skew angle (deg)
ε	Turbulent kinetic energy dissipation rate (m^2/s^3)
κ	Von Karman constant
ω	Specific turbulence dissipation rate (s^{-1})

1 Introduction

An effective conversion of the wind potential in the built environment is becoming a key topic in the field of distributed generation from renewable energy sources. Due to the intrinsic benefits of local integration, focus is particularly given to the use of medium and small wind turbines. In case of rooftop installations or building integration, the conventional siting aspects of the rotor require even more attention, if one wants to correctly exploit the speed up effect due to the interaction between the Atmospheric Boundary Layer (ABL) and the surrounding environment.

The use of computational fluid dynamics (CFD) is well established for large-scale wind farm simulations, while in case of the siting of small machines it is not common to perform dedicated numerical studies for an accurate evaluation of both the local wind conditions and the corresponding energy yield potential. Since the flow conditions above the roof of buildings are different and more complex with respect to an open landscape, the identification of the main characteristics of a suitable site in the built environment is being addressed in current studies. Ledo et al. [1] analyzed the relationship between the wind flow characteristics and the profile of the roofs and they concluded that flat roof is the most favorable shape to yield high power. Abohela et al. [2] analyzed the effects of the roof shape, the wind direction, the building height and the surrounding urban configurations on the wind flow above the roof. The optimum mounting locations were found in correspondence to those roof shapes which allow a central positioning on the roof: velocity increases up to 15% were observed and, consequently, a theoretical enhancement of the energy yield up to 50% could be hypothesized. Similarly, Herrmann et al. [3] showed that an increase up to 29% in the available energy can be gained depending on the configuration of the upwind structure. Some of the authors presented in [4–6] the results of a thorough activity on the estimation of the energetic suitability of both Horizontal Axis Wind Turbines (HAWT) and Vertical Axis Wind Turbines (VAWT) installations in the rooftop of a building in a reference European city. Two dimensional CFD simulations were carried out to characterize the flow field in the rooftop area of an installation building as a function of the proportions of the upwind building and of the average surrounding buildings height. The results were used to calculate the available wind distributions and the yearly energy production of the turbine in view of a more detailed understanding of the performance improvement or detriment related to the specific installation.

Few examples do exist in the technical literature concerning the prediction of the actual energy production of a roof-mounted turbine, mainly due to the remarkable complexity and computational cost of the simulation. Unsteady CFD RANS calculations were presented by Schily and Paraschivoiu [7] and Zanforlin and Letizia [8] for VAWTs, whose geometrical features allow a 2D simplification of the domain. Conversely, fully three-dimensional simulations require an extremely high computational effort and are then suited more to single cases rather than to optimization analyses. For these latter cases, in which multiple configurations have to be investigated, the assessment and tuning of simplified numerical tools for predicting the attended performance in an urban siting may be of major relevance in view of more accurate feasibility analyses. One of the simplest approaches of use to this end is the Actuator Disk Model (ADM), in which the turbine is modelled as a sink of momentum in the main flow. This technique was used by Micallef et al. [9] to assess the power performance of a small HAWT as a function of the height above the roof of a generic cubic building: the results showed that the rotor should be placed at least 3 m above the roof level to conveniently exploit the wind potential.

In this study, a newly-customized version of the Virtual Blade Model (VBM) of the commercial solver ANSYS® FLUENT® was used to evaluate the energetic suitability of a small HAWT installation in the rooftop of a building. With the VBM model, a 3D RANS calculation of the flow field is performed for the domain surrounding the rotor, while the effect of the rotating blades on the fluid is simulated through a body force, calculated by a simplified Blade Element Method (BEM). In previous works [10, 11], the authors compared and validated the results of VBM simulations with both wind tunnel data and fully CFD RANS simulations. The main features of the interaction between turbine wakes, in case of either a single rotor or two rotors, were properly captured by the model. A reliable estimation of the turbine performance was also provided in all the study cases. Conversely, the ADM model did not reveal the same accuracy, especially at higher wind speeds.

The present paper shows the results of a first application of the model to the simulation of a small HAWT placed in the rooftop of a building in a densely built environment. The goal of the analysis was to identify some siting criteria for the rotor and to understand the impact of different locations on the wind energy exploitation. In particular, the effects of both the urban landscape and the turbine positioning were investigated to show the potential of such approach in characterizing the real working condition of the machine. Finally, the prospects of the proposed approach in terms of saving of computational resources in comparison to a full CFD RANS calculation are presented and discussed.

2 Theoretical Approach

2.1 Modelling of the Atmospheric Boundary Layer

The simulation of the ABL flow in urban locations requires an accurate consideration of the effect of the ground roughness on the flow. Indeed, the velocity profile is quite different from the classical log-law profile [12], with a shift of the zero-velocity height (displacement, d) depending on the average height of the surrounding buildings (Fig. 1). A common approach is to use the standard wall functions and to apply the equivalent sand-grain roughness K_s [13] to the wall-adjacent cells to express the effect of terrain on the flow. Blocken et al. [14] found a relationship between K_s and the roughness length z_0:

$$K_s = 9.793\frac{z_0}{C_s}, \tag{1}$$

where C_s is a constant varying between 0 and 1. The near-wall cell size is directly related to the wall roughness, since the distance from the wall of the first mesh node y_p must satisfy the condition of $y_p > K_s$.

The main issue of this approach is to satisfy the y_p constraint both at the ground boundary and on the building walls boundary; for z_0 values in the urban context

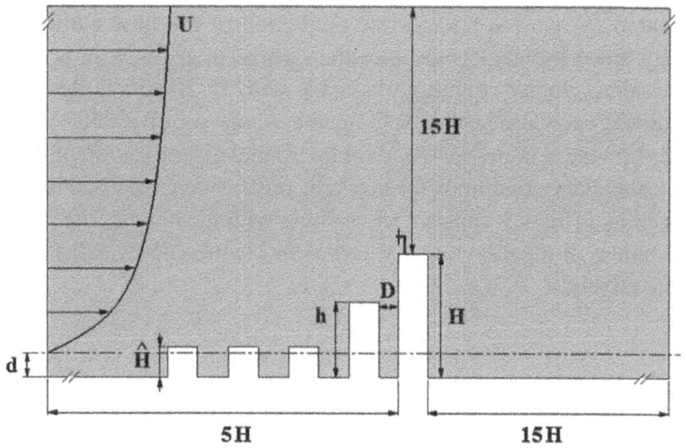

Fig. 1 Side view of the computational domain for the urban landscape

(~1.0 m), the resulting size of the mesh cells at the ground would be excessively large (y_p ~ 30 m) if compared to the grid size on the building walls ($z_0 = 3.0 \times 10^{-4}$ m, y_p ~ 0.01 m). Following the methodology already shown in [4, 5], the roughness length of an open-landscape environment was used at the ground to reduce the first cell height ($z_0 = 3.0 \times 10^{-2}$). An explicit modeling of the roughness elements was then applied by introducing a row of three square blocks in front of the investigated buildings (see Fig. 1) having the function to virtually reproduce the real roughness effects on the flow.

The roughness elements height ($\hat{H}$) corresponds to the average building height of the investigated urban area whereas the number, width and spacing between the elements are chosen in order to preserve the inlet velocity profile.

The steady-state three-dimensional simulations presented in this paper were carried out using the commercial code ANSYS® FLUENT®. The Reynolds-averaged Navier–Stokes (RANS) approach was adopted in a pressure-based formulation, as it is traditionally used for incompressible flows. The operating conditions are standard ambient conditions, i.e. a pressure of 1×10^5 Pa and a temperature of 293 K.

Some of the authors have already assessed and validated the main settings for a proper numerical setup for the simulations of the Atmospheric Boundary Layer in [4, 5] using the OpenFOAM® open-source code. The accuracy of such simulations has been successfully verified by means of experimental wind tunnel measurements of the CEDVAL laboratory [15] on a test case consisting of a single building block (1:200 scale). For the sake of completeness, the main numerical settings are shortly summarized below. The turbulence closure was achieved by means of the *Standard* k-ε model. The use of a ε-based turbulence model is suggested by [2, 3, 7, 9] since its application is recommended in free shear flows, i.e. jet-wake and mixing layers, while ω-based models reveal strong sensitivity on freestream values of both k and ω. On the other hand, ε-based models tend to over-estimate the turbulent kinetic energy in the separated region, but an accurate prediction of the flow within this zone is not necessary since the goal is to place the turbine in zones with no recirculation [9]. The pressure-velocity coupling was made with the SIMPLE algorithm and the residuals convergence was set to 10^{-6} for the whole set of RANS equations. The second order upwind scheme was used for the spatial discretization of the equations.

The inlet boundary conditions for the ABL profile were imposed by the assumption, proposed by [16], of a constant shear stress with the height. The profiles for the velocity, turbulent dissipation rate and turbulent kinetic energy satisfy Eqs. (2), (3) and (4), respectively:

$$V(z) = \frac{u*}{\kappa} \ln \frac{(z-d)}{z_0}, \tag{2}$$

$$\varepsilon(z) = \frac{u*^3}{\kappa z}, \tag{3}$$

$$k(z) = \frac{u*^2}{\sqrt{C_\mu}}, \tag{4}$$

where $u*$ is the friction velocity, κ is the von Karman constant, and C_μ is the turbulence model constant.

2.2 Rotor Model

The Virtual Blade Model (VBM) is a user-defined function implemented in the commercial solver ANSYS® FLUENT®. This add-on was originally conceived to model the aerodynamics of propellers and helicopter rotors and it is still used extensively for these purposes, even in military applications [17]. Its application to HAWTs is, however, quite straightforward [18].

Analogous to the actuator disk approach, the rotor geometry is replaced by a disk-shaped fluid corresponding to the area swept by the turbine blades. In further detail, blades are not explicitly modelled and their effect is simply represented by a body force acting inside the equivalent volume (Fig. 2).

Their time-averaged aerodynamic effects are accounted for through a source term in the momentum equations solved only for the mesh elements of the rotor disk fluid zone. At the beginning of the calculation, the source terms are unknown and they are iteratively updated during the solution, until convergence. This terms are calculated with a classical BEM approach [19], then depending on the chord length, the angle of attack, and the tabulated lift and drag coefficients of the airfoils. In addition, the VBM allows a spanwise discretization of the rotor from the root to the tip in order to compute the momentum sources as a function of twist, chord and airfoil type for each individual segment along the radius. For further details on the numerical implementation of the model, please refer to [20].

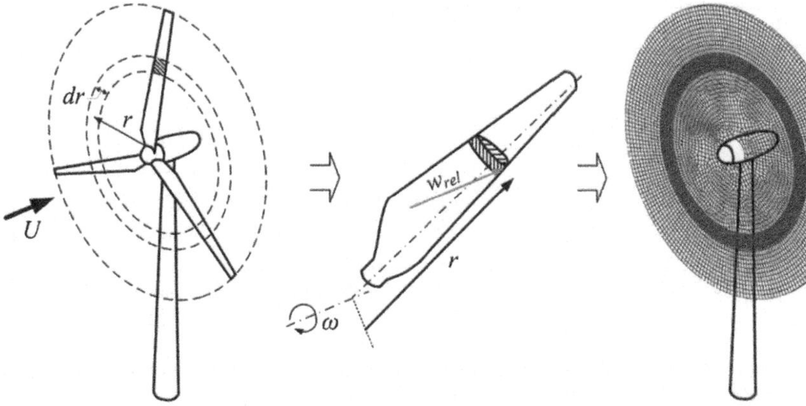

Fig. 2 Schematic representation of the equivalence between a rotor blade element and the corresponding VBM annular area

Moreover, in the present study a customized version of the model was used, in which the stall delay model by Du and Selig [21] was implemented within the VBM. The prediction capability of the VBM was assessed by the authors in Refs. [10, 11] for different test cases including: (1) a single turbine; (2) an array of two turbines with one rotor working in the wake of the other one; (3) an array of two staggered rotors; (4) several configurations of rotors working in yawed-flow. The CFD results showed that the model was able to properly capture the main features of the wake of all the turbines and to provide a reliable estimation of the turbine performance.

For completeness, the only results of the simulations in yawed flow are reported, since they suitably represent the actual operating conditions that can be attended in rooftop installations. Wind tunnel data particularly refer to the tests carried out by Andersen [22] at the NTNU in Norway. The experimental test section was 12 m long, 2.7 m wide and about 1.8 m height. The tested rotor was a 3-blade NREL turbine with a S826 airfoil, a diameter of 0.894 m and a design tip speed ratio (TSR) of 6. Performance and wake measurements were first carried out on a single turbine, which was rotated to induce a yaw angle between the incoming wind and the turbine axis. Then, the effects on the downstream rotor were analyzed.

Figure 3 shows the results of the VBM predictions for the single turbine in yawed flow. Good agreement was appreciated between experiments and simulations for both the maximum *Cp* variation and the wake profiles.

The 2-turbine case is analyzed in Fig. 4, where the power coefficient curves of the downstream rotor as a function of TSR are reported for three relevant yaw angles of the upstream rotor of 0°, 20° and 40°, respectively. Sound agreement was found between experiments and simulations, with a correct description of the *Cp* values, the curve shape and the position of the peak efficiency, confirming the suitability of the model itself for small HAWT siting in the urban environment.

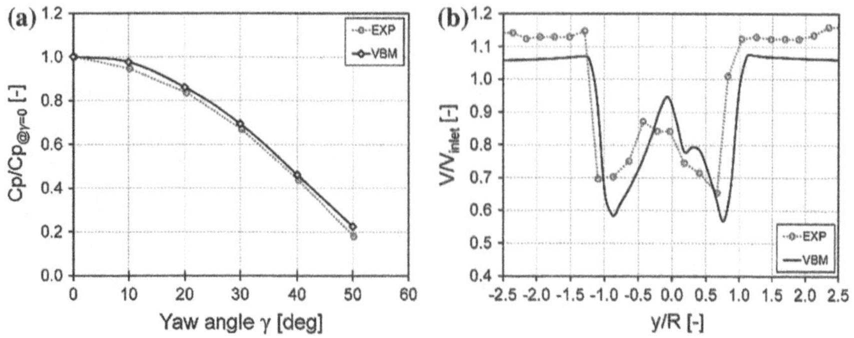

Fig. 3 Test results in yawed flow: **a** reduction of maximum *Cp* as a function of the yaw angle; **b** wake profile description in yawed flow: near wake (1D) for $\gamma = 20°$

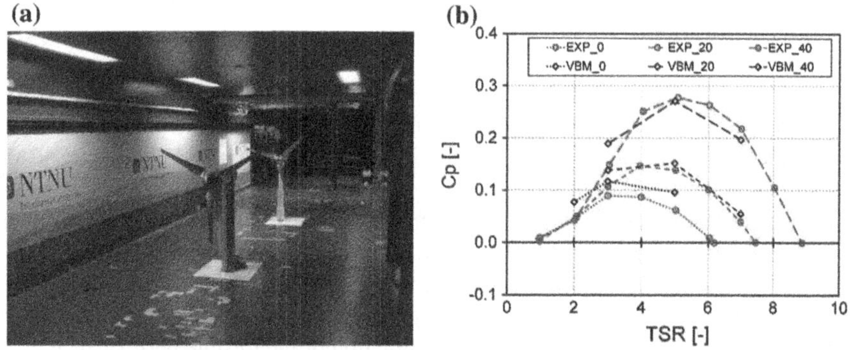

Fig. 4 Test results in yawed flow for two turbines: **a** NTNU wind tunnel; **b** *Cp* versus TSR for the downstream turbine

2.3 Study Cases

According to the literature [2, 12] and to previous authors' experience [4] the installation building (IB) is supposed to be sufficiently higher than the surrounding ones in order to exploit some intense local accelerations of the oncoming flow. The parametric CFD analysis reported by in [4] was carried out to characterize the rooftop flow field as a function of the installation building height *H*, the height *h* of its upwind building (UB) and the distance *D* between the buildings themselves, as shown in Fig. 1. Highest accelerations were observed for an IB height quadruple than the city average height in case of a close distance between IB and UB.

For the present analysis, two urban landscape configurations (namely Environment 1 and *Environment* 2, see Table 1) were selected in order to highlight the influence of the upwind building height on the effective energy yield of a roof-mounted rotor. In both cases, *H* and *D* were fixed to $4\hat{H}$ and $0.5\hat{H}$, respectively. The two values of *h* were chosen to achieve a suitable compromise between high velocity and

Table 1 Urban landscape test cases

Configuration	H	D	h
Environment 1	$4\hat{H}$	$0.5\hat{H}$	$3.25\hat{H}$
Environment 2	$4\hat{H}$	$0.5\hat{H}$	$2.50\hat{H}$

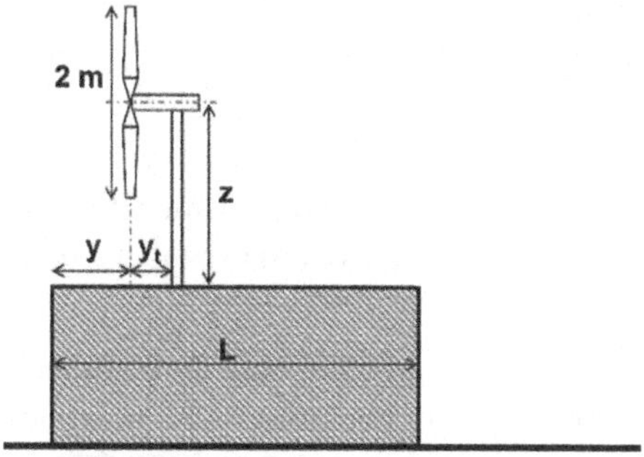

Configuration	y	z
A3	0 m	3 m
A4	0 m	4 m
C3	$y+y_t = L/2$	3 m
C4	$y+y_t = L/2$	4 m
OPT	optimized	optimized

Fig. 5 Analyzed rotor configurations

relatively low skew angle (*Environment 1*) and to exploit the maximum accelerating effect regardless of the skew (*Environment 2*). For all of the buildings, a square base was chosen to maintain the typical proportion of modern tall buildings of city centers.

First, both configurations were simulated without any installed turbine to characterize the flow above the roof of IB.

A study turbine was then designed by means of an aerodynamic scaling of the NREL Phase II, which led to a rotor with a total diameter equal to 2 m [23]. Three different mounting locations were investigated according to the following siting strategies: (1) installation of the turbine in the edge of the building roof with the rotor plane aligned with the building façade. According to Refs. [12, 24] this can be considered the most suitable application in order to exploit the wall effect of the oncoming flow; (2) installation of the turbine with the tower centered in the middle of the building roof, for stability and aesthetical reasons; (3) mounting location optimized through the dedicated CFD simulation of the local flow conditions. In case of centered or aligned positions, two different tower heights were considered since the flow characteristics (i.e. acceleration and turbulence) vary substantially with the height above the roof. Globally, five rotor positions (see Fig. 5) were then simulated for each environment, namely:

Table 2 Roughness properties and the wind characteristics

London data	
Mean building height ($\hat{H}$)	13.6 m
Percentage of the total area occupied by buildings	55%
Roughness length (z_0)	0.29 m
Friction velocity (u*)	0.55 m/s

- *A3* and *A4*: aligned rotor with a 3 m and 4 m tower, respectively;
- *C3* and *C4*: centered rotor with a 3 m and 4 m tower, respectively;
- *OPT*: optimized via CFD.

For the definition of both the parameters needed by Eqs. (1)–(4) and the computational domain dimensions, the roughness properties and the wind characteristics were specified based on a real city data set. In particular, literature data for London city [25, 26] were chosen (Table 2). The displacement value d was calculated by Eq. (5), following the indications of the ESDU [27]:

$$d = \hat{H} - 4.3z_0(1 - A_h) = 13\,\text{m}. \tag{5}$$

2.4 Computational Grid

The dimensions of the computational domain (Fig. 1) were imposed as a function of the tallest building height H, as suggested by [28, 29]. The inlet and outlet boundaries were placed $5H$ upstream and $15H$ downstream of the installation building, respectively. The domain height and width were set to $15H$ and $5H$, respectively.

The definition of the mesh properties in proximity and within the disk replacing the blades was based on a sensitivity analysis carried out on the test case of [22]. Four meshes (labelled from M1 to M4) were tested by progressively doubling the VBM domain elements up to the maximum level of discretization intrinsically allowed by the VBM tool itself (≈8000 elements). The results of the sensitivity analysis were evaluated in terms of the coefficient of determination (R^2) [30] between the velocity wake profiles at different distances from the rotor. From a perusal of Fig. 6, it is apparent that with roughly ≈4000 elements the VBM model is able to guarantee mesh independent results, being $R^2 > 0.98$ for all of the analyzed distances.

The global fluid mesh was obtained by means of an unstructured tetrahedral grid in the whole domain, except for the near-wall regions where an extrusion of prismatic layers was adopted for the boundary layer description. The sizing of the elements was extended from the 2D analyses carried out in [4, 5] to the 3D domain by maintaining the same wall refinement and the same cells expansion ratio. A proper refinement in front of the virtual rotor and in the wake region was also introduced, in order to

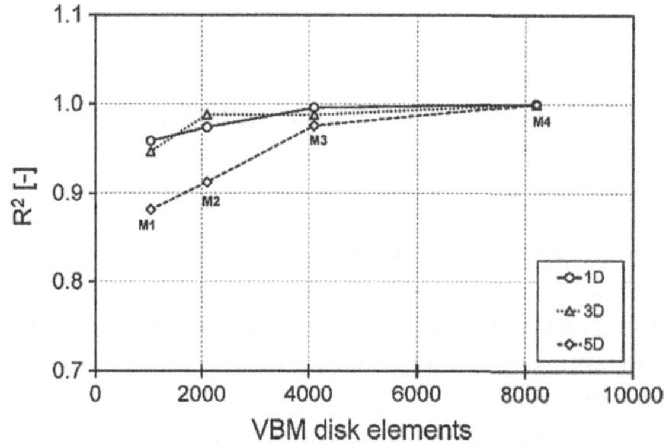

Fig. 6 Mesh sensitivity analysis for the VBM disk

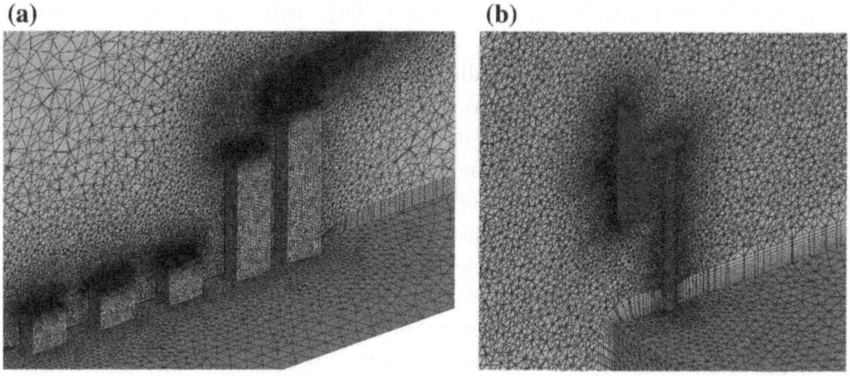

Fig. 7 Computational grid: **a** global view of the mesh of the urban environment **b** detail of the mesh of the VBM disk

accurately resolve the flow field on the rooftop. The resulting grid size was in the range between 6.2×10^6 and 6.6×10^6 mesh cells for all of the rotor positioning. It is worth observing that the VBM approach requires the use of very similar meshes with respect to the simulations of the urban landscape only. Indeed, both *Environment 1* and *Environment 2* require roughly 5.1×10^6 cells for the simulation of the ABL evolution in the urban landscape. Therefore, the introduction of the VBM disk only leads to a 25% increase in the overall cells count.

Figure 7a reports a global view of the surface mesh (grey) for the discretization of the ground and the building walls, and a slice of the mesh (light blue) along the vertical mid-plane to show the mesh details in the fluid region. The increased-density zones on the roofs of each building and the increased-density zone around the VBM mesh, also comprising a considerable part of the domain downstream the IB, are clearly visible. Figure 7b reports a detail view of the mesh of the VBM disk, the turbine tower and the nacelle.

3 Results

3.1 *Wind Turbine Performance Characterization in Yawed Flow*

A preliminary analysis to assess the aerodynamic behavior of the study turbine, both in aligned and yawed flow, was first carried out. The rotor performance was evaluated with a dedicated set of CFD simulations making use of the VBM model within a cylindrical domain representing a hypothetical wind tunnel. For the sake of simplicity, the turbine was simulated only with the rotor disk virtually reproducing the blades, while both the nacelle and the tower were not modeled. The study was carried out for different TSRs with an inflow velocity of 8 m/s at the inlet boundary, considered as a representative value of the average wind speed in the built environment considered further on. To avoid any blockage effect, very large wind tunnel dimensions were set, leading to a test section with a length of 32 m and a radius of 10 m. These dimensions were also sufficient to allow a rotation of the turbine during the tests in order to induce a yaw angle between the oncoming wind and the turbine axis. As recently analyzed by the authors in [10, 11], the VBM model is indeed able to properly model the aerodynamic behavior of the rotors in yawed flow: this feature makes then it particularly promising for all those applications which are characterized by misaligned flows (e.g. the urban environment).

Figure 8 reports the calculated power coefficient curves as a function of the tip-speed ratio and three yaw angles, up to a maximum value of 30°, while the variation of the maximum value of *Cp* as a function of the yaw angle is shown in Fig. 9.

The peak performance was obtained at TSR = 6 for all of the analyzed cases. Upon examination of Fig. 9, one can notice that the yaw conditions do not affect significantly the turbine aerodynamics up to $\gamma = 10°$ compared to the aligned flow case, while differences notably increase for higher angles, following approximately the cosine square of the yaw. Thanks to these preliminary considerations, the optimal

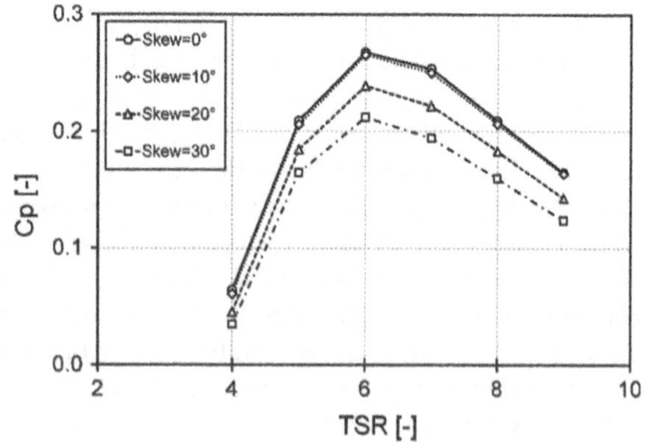

Fig. 8 *Cp* curves as a function of the TSR for different yaw angles

Fig. 9 Reduction of maximum *Cp* as a function of the yaw angle

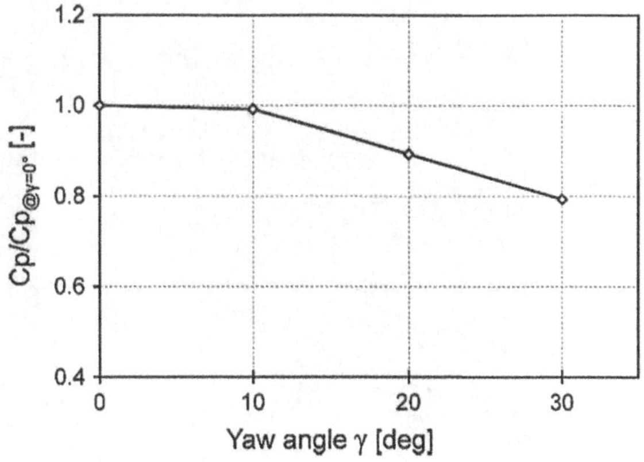

TSR for the turbine simulations in the urban landscape was selected: the value of TSR = 7 was used, being the first stable operating point above the power peak.

3.2 *Built Environment*

CFD simulations of the urban environment were first performed without the turbine to characterize the flow above the roof of the installation building. This preliminary study was needed to evaluate the wind potential in the siting locations defined in Fig. 5 and to identify the locations with locally accelerated flow suitable for the maximization of the energy extraction. Figure 10 reports the velocity streamlines in *Environment 2* to highlight the properties of the three-dimensional flow structures. The flow is not strongly modified by the buildings having the height of the city average. Conversely, high-rise buildings strongly interact with the incoming wind, leading to vortices and regions of accelerated flow. In particular, the aim of the analysis was to investigate the modifications in terms of flow velocity occurring at the rooftop of IB when a different height of UB is considered. The velocity contours on the central vertical plane for both *Environment 1* and *Environment 2* are reported in Fig. 11. The y-component of velocity (V_y) along the streamwise direction was considered, since it is the actual component that leads to the energy extraction by the rotor. The values of V_y are reported in a dimensionless form with respect to the undisturbed y-velocity at the same height of Eq. (2) to clearly highlight the acceleration and deceleration areas for the correct choice of the optimized position.

The five positions on the rooftop where the turbine is thought to be installed are also represented, as defined in Fig. 5. It is apparent that neither the aligned nor the centered rotor configurations are located in the zone of maximum acceleration. Moreover, the area with maximum V_y is not located in the same position for the two environments, testifying that the optimized position actually depends on the height

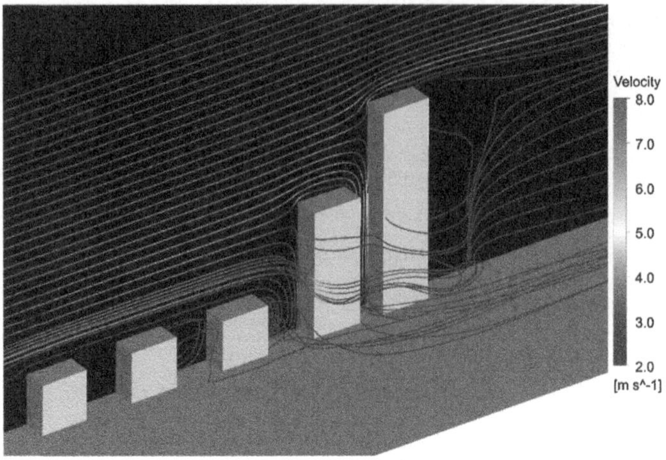

Fig. 10 Velocity streamlines in *Environment 2*

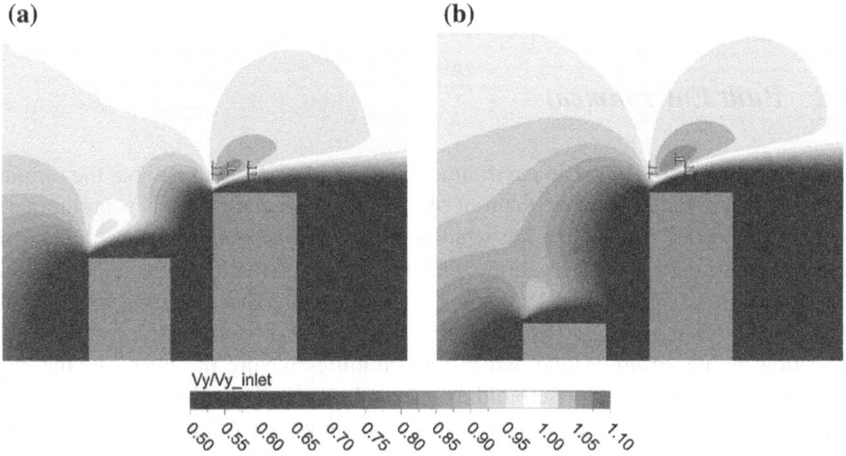

Fig. 11 Dimensionless y-velocity contours: **a** *Environment 1*; **b** *Environment 2*

of UB. In detail, the separation zone has a larger extension for *Environment 2*, due to a different interaction with a lower building, leading to a higher optimal position.

Analogous to previous case, Fig. 12 reports the contours of turbulence intensity on the same vertical plane. According to [1], a turbine should not be exposed to wind with turbulence intensity greater than 16–18% not to affect the operability of the machine. All of the turbine configurations operate within the aforementioned limit, except for case *C3* of *Environment 2*. In general, the closer the turbine to a recirculation zone, the higher the risk of excessive turbulence intensity. Therefore, higher towers are preferred both to avoid turbulence issues and to exploit higher velocities.

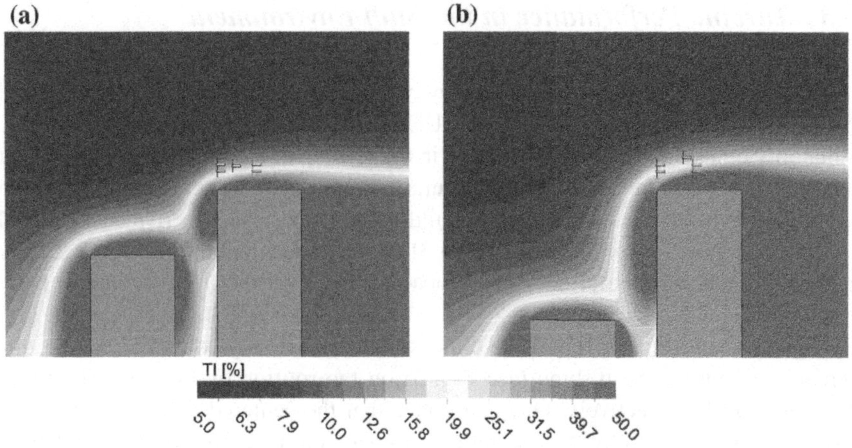

Fig. 12 Turbulence intensity contours: **a** *Environment 1*; **b** *Environment 2*

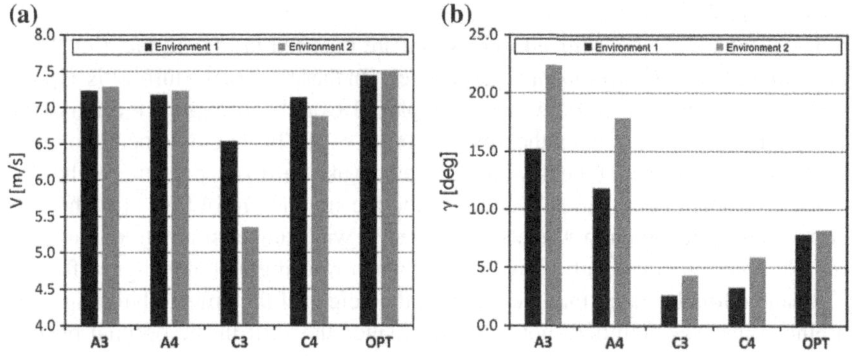

Fig. 13 Comparison of the results between the two environments as a function of the rotor configuration: **a** velocity modulus; **b** skew angle

Finally, flow variations were quantified in terms of velocity modulus and skew angle approaching the turbine, i.e. the angle between the local velocity vector and the wind main direction (perpendicular to the façade). Figure 13 reports the results for the two environments and the five different rotor positions. Aligned positions (i.e. *A3* and *A4*) are characterized by large values of velocity but they also suffer from high inclination angles. Conversely, centered positions (i.e. *C3* and *C4*) allow the turbine to operate with a flow almost normal to the rotor surface, even if the wind speed is lower. The optimal configurations must be then identified as the best compromises between high velocity and moderate skew angles. Moreover, it is worth noticing that the higher accelerations were observed on the top of IB in the *Environment 2*, although this entails higher skew angles compared to *Environment 1*.

In case of aligned configurations, a tilt mounted turbine could also be advantageous to compensate such high values of skew angle and to best exploit the speed up effect.

3.3 Turbine Performance in the Built Environment

After the 3D study of the urban area, a newly-customized version of the VBM model for the commercial solver ANSYS® FLUENT® was used to evaluate the energetic suitability of a small HAWT installation in the rooftop of a building. For the present analysis, the two urban landscape configurations of Table 1 were selected in order to highlight the influence of the upwind building height on the effective energy yield of a roof-mounted rotor. Therefore, the 3D RANS calculations of Sect. 3.2 were replicated including the effect of the rotating blades on the fluid through a body force.

Figure 14 and Fig. 15 report the dimensionless y-velocity contours for four different positions of the turbine (see Fig. 5) on the rooftop of *Environment 1* and *Environment 2*, respectively. One can notice that the centered cases are sited in a slow flow area, caused by the separation occurring on the corner of the installation building. For this reason the wake structure is clearly distinguishable for the aligned cases compared to the centered cases. This discrepancy is more pronounced for the *Environment 2* in Fig. 15.

The results for the optimized rotor siting are reported in Fig. 16 and Fig. 17 for *Environment 1* and *Environment 2*, respectively. In these cases, the turbine is exposed to the maximum flow velocity and it is placed exactly to exploit the maximum acceleration originating from the upwind corner of IB. From the velocity contours, the straightening effect of the flow in the turbine wake due to the presence of the rotor is clearly visible in both cases, if compared to the simulations of Fig. 11 without the turbine. This phenomenon was also observed in wind tunnel simulations, resulting from the steering effect caused by the rotor when working with misaligned flow.

It can be finally stated that the effect of the height of the upwind building in the turbine siting was definitely not negligible, since the shift in the optimal position of the rotor center between the two environments was 1.4 m in height and 1.7 m streamwise.

Finally, the results were quantified in terms of power generated in all of the tested configurations. Figure 18 and Fig. 19 report the power output—expressed in a dimensionless form, with respect to the power of the optimized case—for *Environment 1* and *Environment 2*, respectively. As expected, the maximum power is always provided by the optimized configuration. The performance reduction is quite significant when siting the turbine in arbitrary positions, and the entity of the reduction strongly depends on the skew effect induced by the urban landscape. Indeed, in case of low skewed flows (*Environment 1*) the reduction is between 5–35%, depending on the rotor position, while in case of high skewed flows (*Environment 2*) the reduction increases up to 11–62%. Therefore, the turbine siting is more critical in case of high speed up effect due to the interaction between the ABL and the upwind building.

In addition, it is worth noticing that the reduction is not consistent with the predictions of Fig. 13 for the simulations without turbine, since the y-component of velocity for the *C4* configuration was in the same order of magnitude of *A3* and *A4* configurations for both environments. Conversely, when considering the mutual

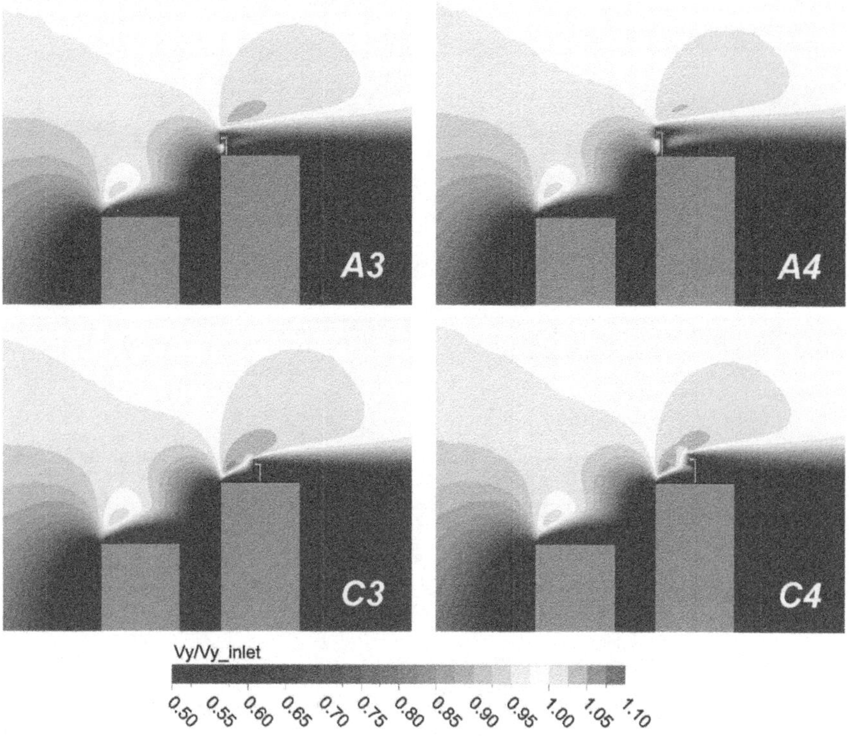

Fig. 14 Dimensionless *y*-velocity contours for the aligned and centered configurations: *Environment 1*

interaction between the turbine and the rooftop flow, the performance predicted for the *C4* configuration decrease, being 8–14% lower than the power in the *A4* position.

As an additional remark, it is worth noticing that the simulations required a small increase in the computational cost with respect to the simulation of the urban landscape only. Indeed, the time needed to achieve convergence was roughly 80% higher, while in case of a full three-dimensional simulation of the rotor geometry the estimated increase would be up to 500%. Based on the above, it can be finally noticed that the Virtual Blade Model confirmed its versatility for this type of analyses and can be then considered worth of consideration as a valuable tool for the turbine siting in the built environment.

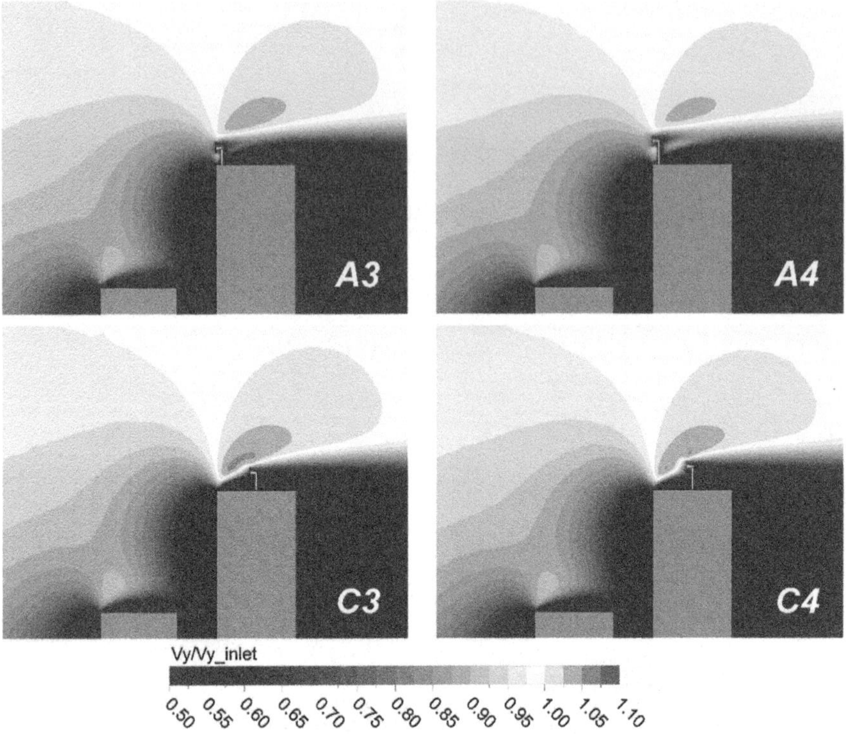

Fig. 15 Dimensionless y-velocity contours for the aligned and centered configurations: *Environment 2*

Fig. 16 Dimensionless y-velocity contours for the optimal configuration: *Environment 1*

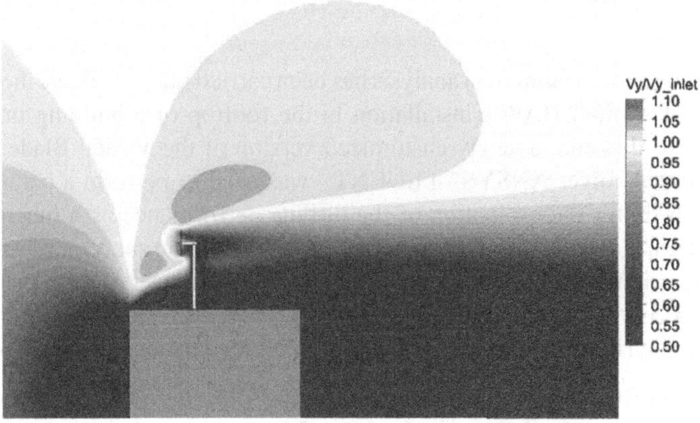

Fig. 17 Dimensionless *y*-velocity contours for the optimal configuration: *Environment 2*

Fig. 18 Power generated in *Environment 1* as a function of the rotor position

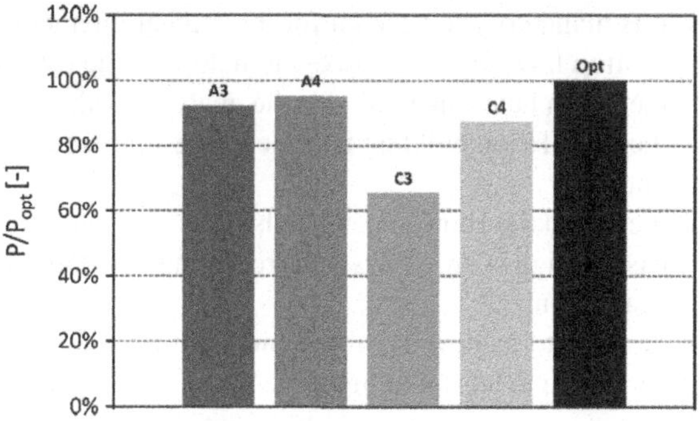

Fig. 19 Power generated in *Environment 2* as a function of the rotor position

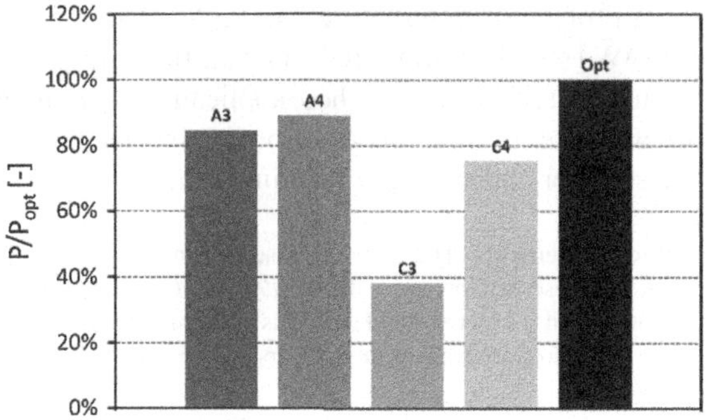

4 Conclusions

In the present study, a numerical analysis has been carried out to evaluate the energetic suitability of a small HAWT installation in the rooftop of a building in an urban landscape. To this end, a newly-customized version of the Virtual Blade Model of the commercial solver ANSYS® FLUENT® was used to perform a parametric 3D RANS analysis for the evaluation of the turbine performance as a function of the urban landscape and the turbine positioning. The goal of the analysis was to identify some siting criteria for the rotor and to understand the impact of different locations on the wind energy exploitation.

The parametric study pointed out some general tendencies, which showed that:

- the choice of the turbine site has a great impact on the power generated by the rotor, due to large variations of the local flow conditions on the rooftop of the installation building. A wrong positioning can lead to a performance drop down to less than 50%;
- it is mandatory to account for the mutual interaction between the turbine and the rooftop flow for an accurate prediction of the rotor performance. A simulation of the urban landscape only for the analysis of the flow conditions on the rooftop of the installation building is indeed not sufficient for correctly estimating the power output;
- the optimal turbine site depends on the surrounding landscape and it is crucial in case of high skew effects, thus requiring a parametric analysis for the performance evaluation;
- in order to achieve notable benefits in the analyzed urban installation, the tower should be at least 4 m high.

In conclusion, the VBM model confirmed its potential in case of non-conventional siting problems, representing a more flexible approach for steady-state simulation of HAWTs with a strong reduction of the requirements in terms of computational resources. Its future use for these applications is thought to potentially provide notable benefits in terms of improved turbine siting, leading to higher energy yields and then to a superior convenience of the investments.

Acknowledgements The activity presented in the paper is part of the research grant assigned to Dr. Francesco Balduzzi by the *Fondazione Cassa di Risparmio di Firenze*, which is sincerely acknowledged for its invaluable effort is sustaining the university research. Thanks are due to Prof. Ennio Antonio Carnevale of the University of Florence for supporting this activity.

References

1. Ledo, L., Kosasih, P.B., Cooper, P.: Roof mounting site analysis for micro-wind turbines. Renew. Energy **36**(5), 1379–1391 (2011)
2. Abohela, I., Hamza, N., Dudek, S.: Effect of roof shape, wind direction, building height and urban configuration on the energy yield and positioning of roof mounted wind turbines. Renew. Energy **50**, 1106–1118 (2013)
3. Herrmann-Priesnitz, B., Calderón-Muñoz, W.R., LeBoeuf, R.: Effects of urban configuration on the wind energy distribution over a building. J. Renew. Sustain. Energy **7**(3), 033106 (2015)
4. Balduzzi, F., Bianchini, A., Ferrari, L.: Microeolic turbines in the built environment: influence of the installation site on the potential energy yield. Renew. Energy **45**, 163–174 (2012)
5. Balduzzi, F., Bianchini, A., Carnevale, E.A., Ferrari, L., Magnani, S.: Feasibility analysis of a Darrieus vertical-axis wind turbine installation in the rooftop of a building. Appl. Energy **97**, 921–929 (2012)
6. Bianchi, S., Bianchini, A., Ferrara, G., Ferrari, L.: Small wind turbines in the built environment: influence of flow inclination on the potential energy yield. J. Turbomach. **136**(4), 041013-041013-8 (2013)
7. Schily, F., Paraschivoiu, M.: CFD Study of a Savonius wind turbine on a rooftop. In: CFDSC, Waterloo, ON (2015)
8. Zanforlin, S., Letizia, S.: Improving the performance of wind turbines in urban environment by integrating the action of a diffuser with the aerodynamics of the rooftops. Energy Procedia **82**, 774–781 (2015)
9. Micallef, D., Sant, T., Ferreira, C.: The influence of a cubic building on a roof mounted wind turbine. In: Science of Making Torque from Wind, TORQUE 2016, Munich (2016)
10. Bianchini, A., Balduzzi, F., Gentiluomo, D., Ferrara, G., Ferrari, L.: Comparative analysis of different numerical techniques to analyze the wake of a wind turbine. In: ASME Turbo Expo 2017, June 26–30, Charlotte, USA (2017)
11. Bianchini, A., Balduzzi, F., Gentiluomo, D., Ferrara, G., Ferrari, L.: Potential of the virtual blade model in the analysis of wind turbine wakes using wind tunnel blind tests. Energy Procedia (2017) (paper in publishing)
12. Mertens, S.: Wind Energy in the Built Environment. Multi-Science, Brentwood (2006)
13. Cebeci, T., Bradshaw, P.: Momentum Transfer in Boundary Layers. Hemisphere Publishing, New York (1977)
14. Blocken, B., Stathopoulos, T., Carmeliet, J.: CFD simulation of the atmospheric boundary layer: wall function problems. Atmos. Environ. **41**(2), 238–252 (2007)
15. Leitl, B., Shatzmann, M.: Compilation of Experimental Data for Validation of Microscale Dispersion Model. CEDVAL, Meteorological Institute, Hamburg University, Hamburg (1998)
16. Richards, P.J., Hoxey, R.P.: Appropriate boundary conditions for computational wind engineering models using the k-ε turbulence model. J. Wind Eng. Ind. Aerodyn. **46**, 145–153 (1993)
17. Laith, Z., Rajagopalan, R.: Navier-Stokes calculations of rotor-airframe interaction in forward flight. J. Am. Helicopter Soc. **40**(2), 57–67 (1995)
18. Javaherchi Mozafari, A.T.: Numerical modeling of tidal turbines: methodology development and potential physical environmental effects. M.Sc. Thesis in Mechanical Engineering, University of Washington (2010)
19. Burton, T., Sharpe, D., Jenkins, N., Bossanyi, E.: Wind Energy Handbook. Wiley, Oxford (2001)
20. Cerisola, A.: Numerical analysis of tidal turbines using virtual blade model and single rotating reference frame. Technical report, University of Washington (2012)
21. Du, Z., Selig, M.S.: A 3-D stall-delay model for horizontal axis wind turbines performance prediction. In: ASME Wind Energy Symposium, January 12–15, Reno, Nevada, paper no AIAA-98-0021 (1998)
22. Andersen, B.: Wake behind a wind turbine operating in yaw. M.Sc. thesis, NTNU, Trondheim, Norway (2013)

23. Tominaga, T., Mochida, A., Yoshie, R., Kataoka, H., Nozu, T., Yoshikawa, M., Shirawasa, T.: AIJ guidelines for practical applications of CFD to pedestrian wind environment around buildings. J. Wind Eng. Ind. Aerodyn. **96**, 1749–1761 (2008)
24. Mertens, S.: The energy yield of roof mounted wind turbines. Wind Eng. **27**(6), 507–517 (2003)
25. Ratti, C., Di Sabatino, S., Caton, F., Britter, R., Brown, M.: Analysis of 3-D urban databases with respect to pollution dispersion for a number of European and American cities. Water Air Soil Pollut. **2**, 459–469 (2002)
26. Martin, C.L., Longley, I.D., Dorsey, J.R., Thomas, J.R., Gallagher, M.W., Nemitz, E.: Ultrafine particle fluxes above four major European cities. Atmos. Environ. **43**, 4714–4721 (2009)
27. Engineering Science Data Unit: Strong Winds in the Atmospheric Boundary Layer, Part 1: Mean-Hourly Wind Speeds. ESDU 82026 with Amendment A and B, London (1984)
28. Franke, J., Hirsch, C., Jensen, A.G., Krüs, H.W., Schatzmann, M., Westbury, P.S., Miles, S.D., Wisse, J.A., Wright, N.G.: Recommendations on the use of CFD in wind engineering. In: International Conference on Urban Wind Engineering and Building Aerodynamics, von Karman Institute, Sint-Genesius-Rode, Belgium (2004)
29. Franke, J., Hellsten, A., Schlünzen, H., Carissimo, B.: Best Practice Guideline for the CFD Simulation of Flows in the Urban Environment. COST Office, Brussels (2007)
30. Mandel, J.: The Statistical Analysis of Experimental Data. Dover Publications, New York (1984)

Numerical Wind Tunnel Tests of an Open Data IPC-VAWT

P. Schito, I. Bayati, M. Belloli, L. Bernini, V. Dossena and A. Zasso

Abstract The evaluation of the performance of a Darrieus Vertical Axis Wind Turbine is performed using an actuator line code. This allows to investigate the aerodynamics of the turbine and the interaction with the boundary conditions. Blade Pitch control, tip-effects modelling and blockage effects in an open-jet wind tunnel test section are studied and investigated. Pitch control allows to enhance the power extraction, tip effects allow to generate a more realistic model, while the modelling of the open-jet test section allows to validate the numerical approach on experimental data intrinsically characterized by not negligible blockage effects.

Keywords CFD · VAWT · Independent Pitch Control · OpenFOAM

1 Introduction

Vertical axis wind turbines (VAWT) have a very complex aerodynamic behavior, characterized by a highly separated and unsteady flow [1]. VAWT can be, at a first stage, separated in Savonius (drag effects) and Darrieus (lift effects) wind turbines. Both turbines are characterized by a highly unsteady flow, therefore a reliable numerical model needs to take into account the actual unsteady flow conditions. In case of Savonius WT, the detailed modelling of the flow field around the turbine is mandatory so that complex numerical models can't be replaced by simplified ones. As a consequence, Savonius WT can be modelled only by CFD approaches, using expensive computations for 3D analysis, that is affordable on a single turbine [2], or 2D analysis for parametric studies on wind farm arrangements [3].

On the other hand, on the basis of a possible simplified modelling of the lift effect, several methodologies are available for the numerical modelling of Darrieus VAWTs [4], from most simple approaches of the Double Multiple Stream Tube (DMST) theory [5], to more complex approaches modelling the flow around the

P. Schito (✉) · I. Bayati · M. Belloli · L. Bernini · V. Dossena · A. Zasso
Politecnico di Milano, via La Masa 1, 20156 Milan, Italy
e-mail: paolo.schito@polimi.it

L. Battisti and M. Ricci (eds.), *Wind Energy Exploitation in Urban Environment*,
Green Energy and Technology, https://doi.org/10.1007/978-3-319-74944-0_8

turbine itself. These approaches consist in vortex methods [6], panel methods [7] and Computational Fluid Dynamics (CFD) models. Historically, VAWT made use of methods originally developed for Horizontal Axis Wind Turbines (HAWT), although in case of HAWT simpler approaches can be implemented due to the symmetry around the rotor axis. The development of High Performance Computer (HPC) facilities lead to the possibility to resolve very complex flows, modelling with increasing detail the turbulent structures of the flow. The most complex models can describe with high accuracy the blade-flow interaction, numerically modeling the blade boundary layer during its rotation, but current HPC facilities are not sufficient to perform parametrical studies on the performance of the turbine itself. At this purpose a simplified approach shows nowadays the best modelling performances, relying on the sectional airfoil aerodynamics of the blades generalized through a hybrid CFD approach to the 3D entire turbine. The hybrid approach allows to insert in the CFD equations the aerodynamic actions of the blades directly in terms of momentum source terms, using tabulated airfoil data to compute the lift and drag forces as a function of the airfoil section relative flow. This technique, referred as High-Fidelity modelling of HAWT is currently performed using an actuator line modelling (ALM) of the aerodynamic forces acting on the blades formerly developed by Sorensen and Shen [8]. Recent implementation of ALM to VAWT if described in [9]. The ALM technique allows finally a straight forward implementation within a CFD modelling of wind farm wakes and atmospheric boundary layer effects as shown in [10].

An original contribution to the ALM technique has been recently developed at Politecnico di Milano (POLIMI) in the OpenFOAM framework through the Effective Velocity Model (EVM) improving the estimate of the actual airfoil section relative flow and consequent aerodynamic forces [11–13]. At the same time an original implementation of this model to VAWT has been developed at POLIMI accounting for implementation of the blades Independent Pitch Control feature (IPC) [14, 15], complex boundary conditions as the wind tunnel open jet flow [16, 17] and corrections for the complex 3D effects of the VAWT wake and the tip effects of the turbine [16, 17].

The VAWT turbine that studied in the present paper is a prototype three bladed H-Darrieus rotor, with Independent Pitch Control (IPC), developed at POLIMI within the VODCA project and fully referenced in [18]. The large open jet configuration of POLIMI Wind Tunnel [19] allowed for the construction of a large size prototype, achieving large Reynolds numbers and providing valuable data finalized at experimental validation of different numerical modelling approaches. Key purpose of the high fidelity ALM developed at POLIMI and implemented on VAWT is to allow the most effective use of the experimental prototype data, accounting for the actual testing conditions inclusive of unavoidable wind tunnel blockage effects and overall 3D flow features, e.g. making the prototype results valuable for validation of reduced order numerical models.

The present paper shortly describes the POLIMI ALM implemented on VAWT, investigating the effect of different conditions, such as IPC strategies, tip effects modelling and boundary conditions effects, comparing the VAWT performances expected in open air testing with those expected in POLIMI open-jet wind tunnel testing. IPC predictive features of the numerical model have been also tested for future application on improving the energy harvesting capabilities of the Darrieus VAWT, as well as possible structural load dynamics reduction and implementation of possible control-driven self–starting capabilities.

2 VAWT Numerical Model

The H-Darrieus VAWT available at Politecnico di Milano is extensively described in [18], since it is used as a benchmark for the VODCA project. The geometry is reported in Fig. 1, where the main dimension is reported. The blade airfoil is a NACA0021, with a radius of 1 m and a total height of 2.7 m. The turbine is equipped with a main shaft motor and with three independent blade motors for the actuation of the pitch for each blade. The large scale model allows to test the wind turbine with a Reynolds number equal to the full–scale one and to realize all mechanical and actuation components with a size that can be implemented in a commercial wind turbine concept.

The turbine will be tested in the open-jet test section of the Politecnico di Milano Wind Tunnel reported in Fig. 2. The characteristics of this section is a size of 4×3.8 m, a maximum wind speed of 55 m/s and a turbulence $I < 0.1\%$.

It is evident that the dimension of the turbine in relation with the size of the wind tunnel test section arises large blockage effects: to limit the interference between the turbine and the wind tunnel test section, it is expected to test the wind turbine in an open-jet condition. Blockage effects will not be negligible when comparing the VAWT performances with respect to an open–field condition, therefore the need for a tool for the assessment of the blockage effects is needed. If blockage effect cannot be estimated the data obtained by wind tunnel experimental tests will not be useful for the validation of numerical models that cannot take into account of the

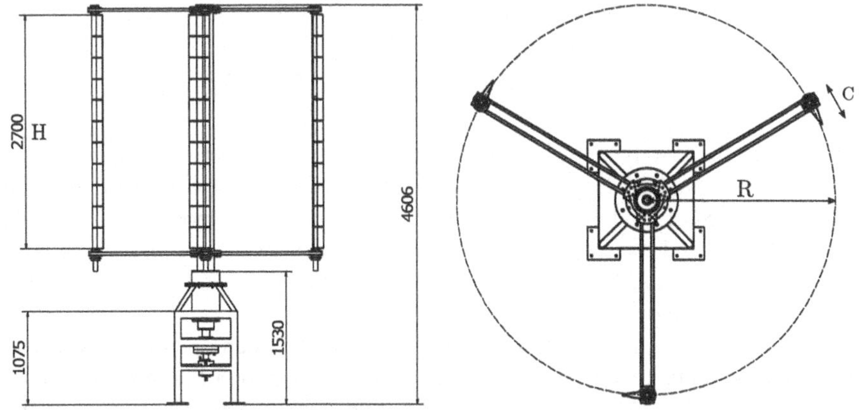

Fig. 1 H-Darrieus VAWT geometry and main dimension

Fig. 2 Politecnico di Milano wind tunnel drawing: when the yellow section is removed the tunnel works with an open-jet condition

wind tunnel testing conditions or with respect to the performances attended for a open–field operation of the turbine. The numerical tool that allows to estimate the wind turbine performance in an open–jet condition is an actuator line modelling, since it is possible to compute the interaction of the flow with both the turbine and the boundary conditions.

The numerical model of the wind turbine is realized in the OpenFOAM framework, implementing the aerodynamic forces in the Detached Eddy Simulation (DES) equations. The VAWT numerical model was firstly developed for HAWTs by Schito [11] in his PhD thesis; successively it was improved by two master students in [12]. Successively it was applied to VAWT by Schito and Zasso [13] in its master thesis; the open-jet applications and tip effects were studied by Ferlini [17]. Momentum equation can be generally written as:

$$\frac{\delta \rho \boldsymbol{u}}{\delta t} + \nabla \cdot (\rho \boldsymbol{u}\boldsymbol{u}) - \nabla \cdot (\mu \nabla \boldsymbol{u}) + \boldsymbol{f} = -\nabla \mathrm{p}$$

where **f** are the aerodynamic forces computed using the Effective Velocity Model (EVM). The standard Improved Delayed DES solver of OpenFOAM, using a Smagorinsky SGS model and a Spalart–Allmaras near wall modelling, has been modified to solve the aerodynamic forces of the VAWT. Each time-step the aerodynamic forces are computed according to the actual flow field and introduced in the momentum equations.

Two different meshes are realized: one that reproduces an open-field condition and one representing the open-jet test section of Politecnico di Milano wind tunnel (see Fig. 3). The mesh is mainly structured, with some trimmed cells close to the tunnel walls. The open-jet test section of the wind tunnel is modeled in the CFD

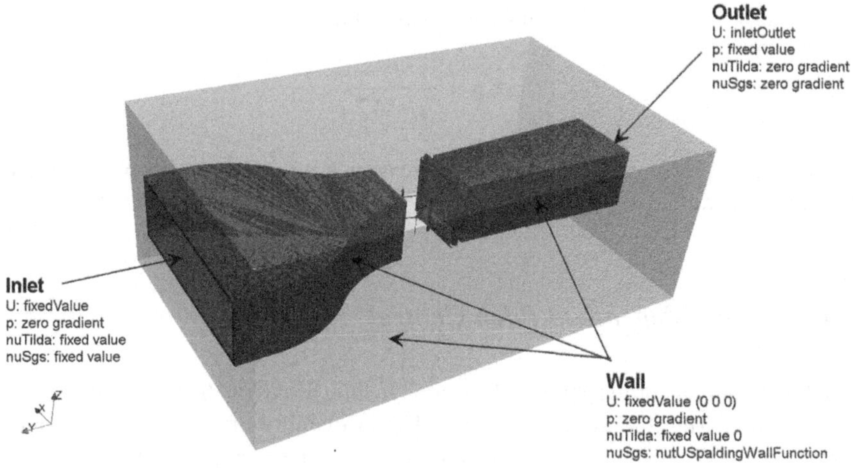

Fig. 3 Open-jet test section geometry modeling in OpenFOAM

model, using the OpenFOAM tools for meshing, allowing to consider the blockage effect of the turbine itself [16, 17].

2.1 Effective Velocity Model

The effective velocity model has been developed at Politecnico di Milano to correctly model the aerodynamic interaction between an airfoil and the air using an actuator line method [11, 12]. Standard actuator line modelling uses the circulation theory that relies on the center of pressure location, however the discretization errors introduced by the CFD mesh introduces some errors that may be limited when using the EVM model. The detailed description may be found in [13], but the main idea is to relate the flow upstream an airfoil with the reference wind speed and angle of attack of the airfoil. The flow approaching the airfoil is affected by the airfoil itself (change in wind speed and direction), but it is possible to define a relation between the lift and drag coefficient of the wing profile and the induction of the airfoil on the flow.

3 Results

Performances of the wind turbine are presented evaluating different indicators: the power coefficient as a function of Tip Speed Ratio (TSR), the tangential force as a function of blade azimuthal position and the flow field around the turbine. The sensitivity of the turbine performance is compared against the open-field condition.

The turbine power coefficient C_P is computed as:

$$C_P = \frac{P}{\frac{1}{2}\rho A V^3}$$

where P is the power generated, ρ is the air density, A is the frontal area of the turbine, V is the undisturbed wind speed.

3.1 Blade Independent Pitch Control

The forces and the torque generated by aerodynamic forces on the turbine show large oscillations during the turbine rotations. The availability of IPC allows to exploit strategies for the reduction of the load oscillations, the increase of the power production and to investigate possible pitch actuation strategies to obtain self–starting capabilities to the turbine. The effect of IPC on the turbine performance has been evaluated by applying a sinusoidal wave oscillation to the blade, in analogy with a cyclic pitch control, but more advanced and effective control laws are under investigation. The effect of cyclic pitch variations with an amplitude of 5° and a period of 1 turbine revolution can be observed in Fig. 4: the performance of the wind turbine with fixed blades is compared against the IPC. The performance of a turbine with IPC is higher than the fixed blades for all turbine working conditions, with lower improvements towards the regions where extreme TSRs are considered. The variation of the blade pitch angle allows to enhance the production of power of the turbine for the main blade azimuthal angles, increasing the tangential force during the turbine rotation.

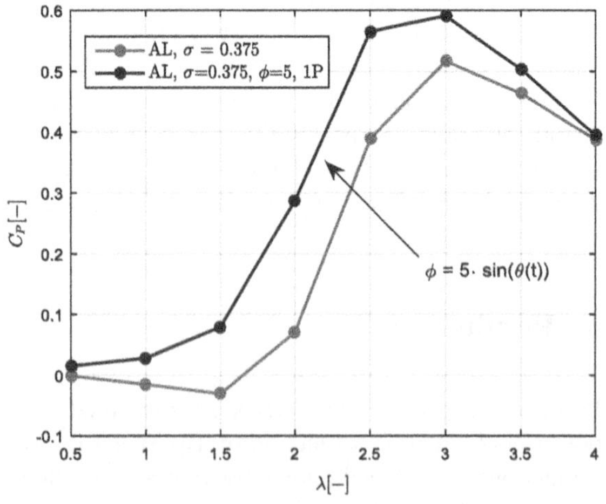

Fig. 4 IPC effect on Darrieus VAWT power coefficient

3.2 Tip Effects

To include tip effects some theories have been recovered from HAWT modelling. The classic tip effects corrections that are adopted are the Prandtl correction and the Glauert correction [17].

Prandtl correction is expressed as:

$$F = \frac{2}{\pi}\cos^{-1}\left[\exp\left(-\frac{B\,(R-r)\,\sqrt{1-\lambda^2}}{2R}\right)\right]$$

while Glauert correction is written as:

$$F = \frac{2}{\pi}\cos^{-1}\left[\exp\left(-\frac{B\,(R-r)}{2r\,\sin\varphi}\right)\right]$$

where B is the number of blades, R the length of the blade, r the distance from the blade center, λ is the TSR, φ is the airfoil angle of attack.

As can be seen in Fig. 5, the actuator line model is not able to consider tip effect, showing a small decrease of the lift force towards the blade tips, mainly due to the deflection of the flow imputed to the drag force of the turbine, rather than by the tip vortices. When computing the tip effects using Prandtl or Glauert correction, the effects are much more localized and noticeable.

The effect on the power coefficient of the turbine is shown in Fig. 6, where the effects are more evident in the high C_P region, rather than in the low efficiency region.

3.3 Boundary Condition Effects

Tests on the Vodca VAWT available at Politecnico di Milano will be conducted in the open-jet test section, where blockage effects cannot be ignored. To evaluate the

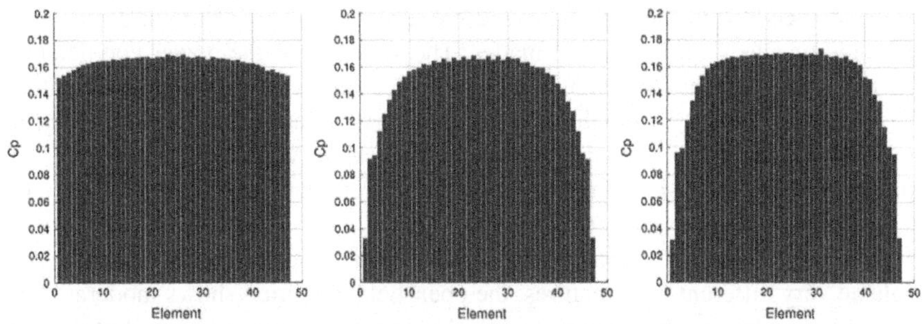

Fig. 5 Lift force developed along a blade: neglecting tip effects (left), considering Prandtl correction (center) and considering Glauert correction (right)

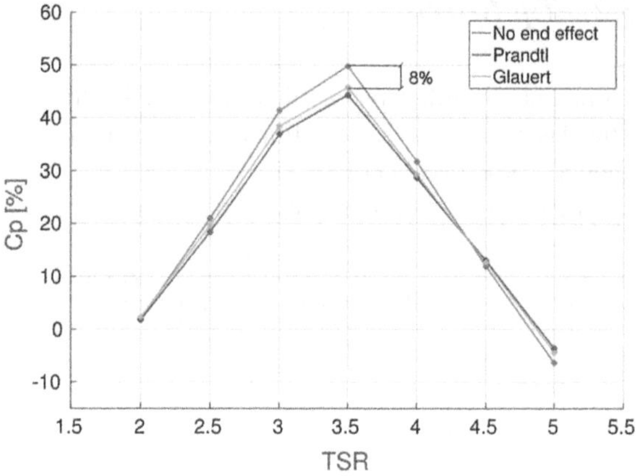

Fig. 6 Power coefficient of the VAWT including tip effects

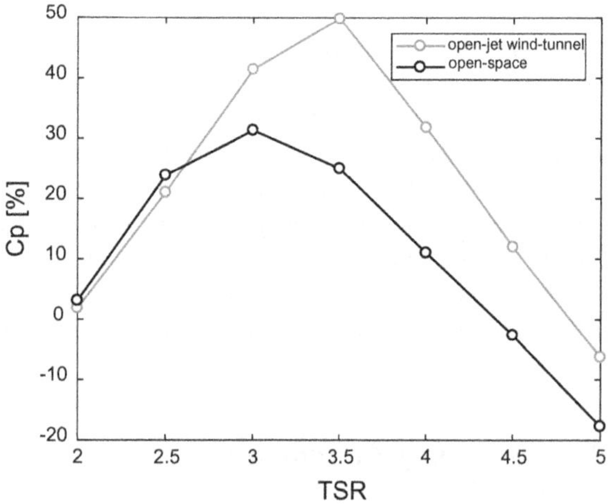

Fig. 7 Power coefficient of the VAWT rotating in an open-field condition and in the open-jet

blockage effects, the comparison of the CFD analysis for the open-field condition and for the open-jet condition has been evaluated [17]. The blockage is mainly affecting the wind speed up in the region around the turbine, affecting the wind speed and the angle of attack of the turbine blades. The variation of the power coefficient is reported in Fig. 7, showing significant differences especially for larger C_P values. This is mainly related also to the VAWT global drag value, since the wind speed-up close to the turbine blades is different in the two cases. Very large values of C_P are computed for the open-jet wind turbine, since the reference wind speed is evaluated as the flow rate in the wind tunnel section.

Looking to the flow characteristics close to the wind turbine, it is possible to evaluate very different flow features: the open-field condition shows moderate flow accelerations close to the wind turbine wake, and wake zone with low momentum, while the open-jet test section shows a larger flow acceleration close to the wake, and higher wind speed in the wake region, suggesting that the wind turbine kinematics

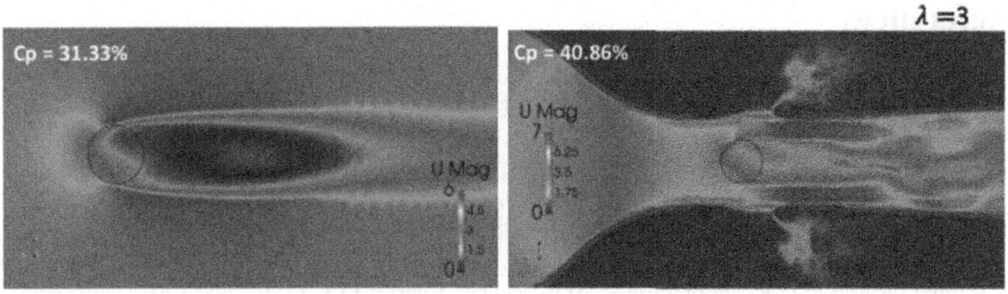

Fig. 8 Flow field on a plane perpendicular to the turbine axis in the open-field condition (left) and the open-jet simulation (right). Turbine is working with a TSR = 3

has blade incidence angles very different in the two cases. This result is evident when looking to the power coefficient values that show large discrepancies (Fig. 8).

4 Conclusions

An actuator line model of a Darrieus Vertical Axis Wind Turbine has been realized in the open–source framework of OpenFOAM. This tool showed appreciable capability to reproduce the complex aerodynamics of a VAWT, allowing also an efficient implementation of the blades Independent Pitch Control (IPC) feature realized on the experimental prototype available at POLIMI. The actual 3D flow features at the blade tip has been captured through implementation of Prandtl and Glauert correction methods, and the specific boundary conditions of the prototype wind tunnel experiment were considered through a detailed modelling of the actual open jet test section flow configuration and associated complex blockage features. The complete numerical model finally implemented was able to offer the appropriate numerical environment for an adequate use of the experimental data provided by the prototype VAWT at experimental vs numerical validation purposes.

The first simulations performed showed the relevant role played by the actual open jet test section features influencing at a great extent the turbine performance. It was shown that for the experimental validation of reduced order VAWT models, unable to reproduce the actual experiment open jet blockage, the speed–up effect due to the experimental testing boundary conditions must be considered.

The VAWT actuator line modelling technique implemented showed high fidelity in modelling many different turbine working conditions, so that offering a valuable tool for verifying and enhancing the reduced order VAWT models, on the basis of a correct reference to the experimental validation prototype data.

References

1. Paraschivoiu, I.: Wind Turbine Design with Emphasis on Darrieus Concept. Polytechnic International Press (2002)
2. Ferrari, G., Federici, D., Schito, P., Inzoli, F., Mereu, R.: CFD study of Savonius wind turbine: 3D model validation and parametric analysis. Renew. Energy **105**, 722–734 (2017). https://doi.org/10.1016/j.renene.2016.12.077
3. Mereu, R., Federici, D., Ferrari, G., Schito, P., Inzoli, F.: Parametric numerical study of Savonius wind turbine interaction in a linear array. Renew. Energy **113**, 1320–1332 (2017). https://doi.org/10.1016/j.renene.2017.06.094
4. Simão Ferreira, C., Madsen, H.A., Barone, M., Roscher, B., Deglaire, P., Arduin, I.: Comparison of aerodynamic models for vertical axis wind turbines. J. Phys. Conf. Ser. **524**(1) (2014)
5. Strickland, J.: The Darrieus turbine—a performance prediction model using multiple streamtube. Technical Report SAND75-0431, SANDIA National Laboratories (1975)
6. Murray, J.C., Barone, M.: The development of CACTUS: a wind and marine turbine performance simulation code. In: 49th AIAA Aerospace Sciences Meeting AIAA 2011-147, Orlando, FL (2011)
7. Simão Ferreira, C.: The near wake of the VAWT: 2D and 3D views of the VAWT aerodynamics. Ph.D. thesis, Delft University of Technology (2009). ISBN/EAN:978-90-76468-14-3
8. Sorensen, N.N., Shen, W.Z.: Numerical modelling of wind turbines wakes. J. Fluids Eng. **124**(2), 393–399 (2002)
9. Shamsoddin, S., Porté-Agel, F.: Large eddy simulation of vertical axis wind turbine wakes. Energies **7**(2), 890–912 (2014)
10. Shamsoddin, S., Porté-Agel, F.: A large-eddy simulation study of vertical axis wind turbine wakes in the atmospheric boundary layer. Energies **9**, 366 (2016)
11. Schito, P.: Large eddy simulation of wind turbines: interaction with turbolent flow. Doctoral Dissertation, Politecnico di Milano (2011)
12. Bernini, L., Caccialanza, M.M.: Development of the effective velocity model for wind turbines aerodynamics numerical simulation through an actuator line approach. Master thesis in Mechanical Engineering. Politecnico di Milano (2014)
13. Schito, P., Zasso, A.: Actuator forces in CFD: RANS and les modeling in OpenFOAM. J. Phys. Conf. Ser. **524**(1) (2014). https://doi.org/10.1088/1742-6596/524/1/012160
14. Caglioni, M.: Actuator line method based on the effective velocity model for CFD simulations of vertical axis wind turbines. Master thesis in Mechanical Engineering. Politecnico di Milano (2015)
15. Caglioni, M., Bernini, L., Bayati, I., Schito, P., Zasso, A.: Development of the actuator line method based on the effective velocity model for CFD simulations of vertical-axis wind turbines. In: Symposium on OpenFOAM in Wind Energy, Milano, Italy (2015)
16. Ferlini, M., Bernini, L., Schito, P., Zasso, A.: CFD model of a vertical-axis wind turbine in open-jet wind-tunnel. In: Symposium on OpenFOAM in Wind Energy, Pamplona, Spain (2017)
17. Ferlini, M.: CFD model of a vertical-axis wind turbine in open-jet wind-tunnel. Master thesis in Mechanical Engineering. Politecnico di Milano (2017)
18. Bayati, I., Foletti, S., Tarsitano, D., Belloli, M.: A reference open data vertical axis wind turbine, with individual pitch control, for code validation purposes. Renew. Energy (2017). https://doi.org/10.1016/j.renene.2017.08.090
19. Politecnico di Milano wind tunnel. www.windtunnel.polimi.it

Wake Measurements of Small-Scale Vertical Axis Wind Turbines at Politecnico Di Milano: A Critical Review

G. Persico, V. Dossena and A. Zasso

Abstract In the last ten years, four measurement campaigns were performed at Politecnico di Milano on two Darrieus Vertical Axis Wind Turbines (VAWT) for micro-generation of different architecture (H-shaped vs troposkien), but sharing the blade number (3), the blade profile (NACA 0021), and the swept area (1.5 m^2). The experiments, carried out in the large-scale wind tunnel of Politecnico di Milano, included detailed wake measurements. This paper presents a review of the research activities related to velocity and turbulence measurements in the wake, proposing an analysis of both the technical aspects and the scientific outcomes of the investigation. In particular, the wakes of these turbines were measured on several surfaces downstream of the rotors for different tip speed ratios and different Reynolds numbers, searching for corresponding conditions between the two rotors. The paper first presents the technical issues involved in measuring the flow velocity in the wake of VAWT rotors with intrusive techniques such as hot wire anemometers and pressure probes. The second part of the paper proposes a comprehensive analysis of the wakes shed by the tested models. The wakes appear asymmetric and roughly follow the shape of the rotor, their width and velocity deficit being strongly dependent on the tip speed ratio. Flow angle measurements show the onset of large-scale tip vortices, for both the H-shape and the troposkien rotors, even though resulting from different aerodynamic mechanisms in the two architectures. A discussion on the impact of the wake features on the implementation of VAWTs in the urban environment concludes the paper.

Keywords VAWT · H-shaped and troposkien rotors · Wind tunnel testing
Turbine wakes · Hot-wire anemometry

G. Persico (✉) · V. Dossena
LFM - Laboratorio Di Fluidodinamica Delle Macchine, Dipartimento Di Energia, Politecnico Di Milano, Via Lambruschini 4, 20158 Milan, Italy
e-mail: giacomo.persico@polimi.it

A. Zasso
Dipartimento Di Meccanica, Politecnico Di Milano, Via La Masa 1, 20156 Milan, Italy

L. Battisti and M. Ricci (eds.), *Wind Energy Exploitation in Urban Environment*, Green Energy and Technology, https://doi.org/10.1007/978-3-319-74944-0_9

1 Introduction

Harvesting the energy made available by the wind in the urban environment is a challenging task that still needs proper demonstration. Several concepts have been proposed in Literature (see, for example, [1–4]) leading to the so-called Building-Augmented Turbines (BAWT) and suggesting that, in the urban environment, the vertical axis wind turbine (VAWT) might be competitive against the horizontal axis wind turbine (HAWT), that nowadays dominates the market of wind energy conversion. In presence of buildings, that interacts in a complex way with the atmospheric boundary layer, the insensitivity to yaw, thanks to the peculiar shape of the VAWT rotors, and the lower acoustic pollution, provided by the lower optimal tip speed ratio (TSR), make the VAWTs competitive with respect to their horizontal counterparts, despite their inherent flow complexity and fatigue loads. Among the several concepts proposed for VAWTs, such as drag-driven machines and hybrid drag-lift Savonius turbine, the most promising configuration is the lift-driven Darrieus turbine, which provides the highest performance that can be made even comparable to the one of the HAWT. Also, for VAWTs simple models are available in Literature to correct their performance and loads in skewed flow [5].

Among the challenges that the design of BAWT has to face, the optimal integration of the rotors within the urban environment is one of the most critical and needs to be pursued with Computational Fluid-Dynamic (CFD) models including the atmospheric boundary layer, the buildings, and the turbine. The ideal simulation tool for such complex configuration (the unique that leads to a reasonable computational cost), is the so-called actuator-line CFD model, proposed in [6] for VAWT and further developed in [7, 8], in which the blades of the turbine are replaced by rotating volume forces. These forces are provided through the profile polars.

Due to the significant modeling effort required by this kind of simulation tool, a proper validation of the computational model is required, for what pertains to both the rotor performance and the wake.

Unfortunately, the low industrial penetration of the VAWT technology has implied a still limited knowledge on the turbine aerodynamics, especially regarding the wake shed by these turbines. Therefore, a significant need of experimental data on the wake shed by VAWT is crucial to investigate the technical feasibility of the use of this technology in the urban environment.

These considerations acted as motivation for a wide experimental campaign on micro vertical axis wind turbines, launched in 2009 as a collaboration between Politecnico di Milano (IT) and Università di Trento (IT). The study was performed on two Darrieus rotors (an H-shaped rotor and a troposkien one), that share the blade profile (NACA 0021), the number of blades (3) and the swept area. The turbines were designed and manufactured under the supervision of the turbomachinery Research Group of the Università di Trento (IT). The aerodynamics of the turbines were investigated in controlled conditions, by resorting to wind tunnel measurements performed in the large-scale wind tunnel of Politecnico di Milano. Four test campaigns were carried out between 2009 and 2016. Measurements included both aerodynamic torque

Table 1 Main geometrical features of the tested rotors

Rotor type	H-shaped	Troposkien
Swept area, A (m^2)	1.5	1.5
Blade profile	NACA 0021	NACA 0021
Blade chord, c (mm)	85	85
Blade number, N (−)	3	3
Rotor equatorial diameter, D (m)	1.030	1.510
Rotor height, h (m)	1.457	1.510
Blade length, L (m)	1.457	2.233
Rotor solidity, σ (−)	0.25	0.38

and thrust recorded through a high precision test bench on the rotor, and also considered the characterization of the turbine wakes by means of time-resolved velocity measurements by traversing multiple hot-wire and pressure probes.

In the present paper, focus is made on the wake profiles and their unsteady evolution, with the aim of reviewing the main indications inferred by the experiments, identifying the similarities and the differences between the wake shed by the two machines.

2 Turbine Models

The section presents the two VAWT models considered in this research. Table 1 summarizes the main geometrical features of the two rotors. As recalled in the Introduction section, the turbines are one of H-shape and another one of troposkien architecture; both the rotors have 3 blades which implement the same symmetrical profile NACA 0021 with chord equal to 85 mm; the rotors also share the same swept area (1.5 m^2). The rotor solidity σ appearing in Table 1 is computed according to the equation $\sigma = N\, c\, L/A$. Figure 1 depicts the images of the rotors.

3 Experimental Set-up

3.1 The Large-Scale Polimi Wind Tunnel

The large scale wind tunnel of the Politecnico di Milano (IT), shown in Fig. 2, consists of a closed-loop facility in which the air is circulated by two arrays of multiple fans working in parallel. The facility is a double-tunnel system; the first tunnel has a larger cross-section (16 m width and 3.84 m height) and operates at low wind speed (up to

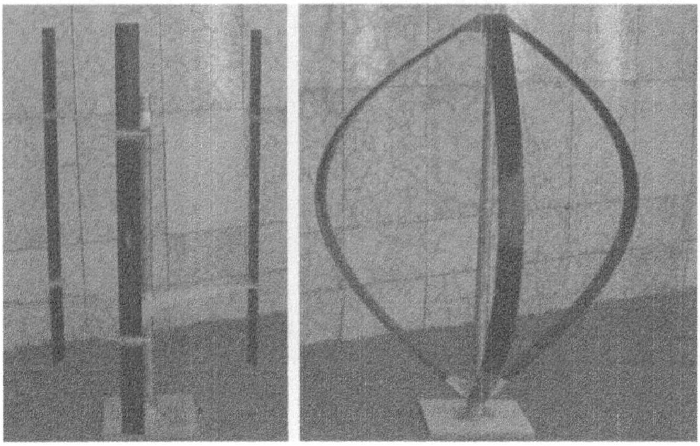

Fig. 1 The H-shaped rotor (left) and the troposkien-shaped rotor (right), both with NACA 0021 profile

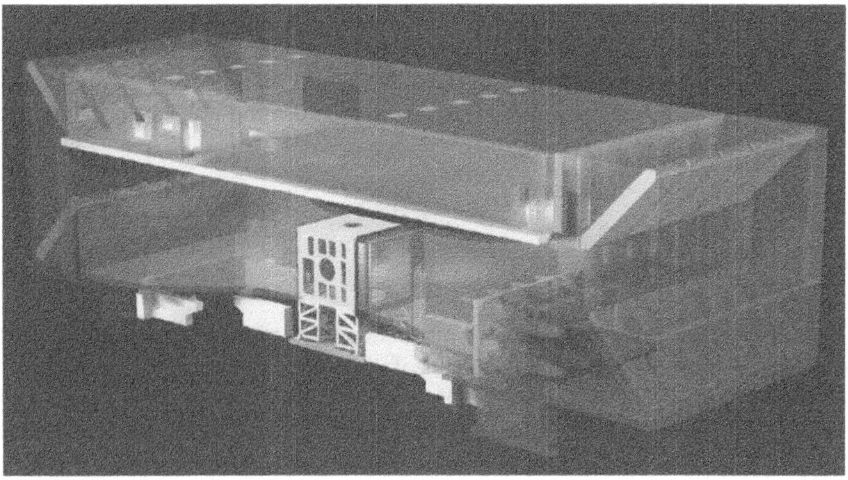

Fig. 2 Schematic view of the large scale wind tunnel of the Politecnico di Milano (IT)

16 m/s); the second one operates at high-speed (up to 55 m/s) and has 4.00 m width, 3.84 m height and 6.00 m length. The turbines were tested in this latter section.

High flow quality in the high-speed section is assured by a combination of honeycombs and anti-turbulence screens, leading to very low a freestream turbulence intensity (<1%) at the rotor test section.

For the present test campaign, the wind tunnel was operated using a "free jet" (open channel) configuration, obtained by removing the test room and positioning the rotor directly facing the upstream tunnel and at the center of the jet. Comparative tests were also performed in closed channel configuration, in which the two turbines have a solid blockage (defined as swept area divided by the total area of the cross section) of 10%.

Throughout the study, an estimate has been done for the free-jet blockage, as a few studies exist in Literature on this topic [9, 10]. The procedure proposed in [10] has been applied to both the turbines in the present configuration (see [11] for more details); it indicates that the blockage correction ratio for the present cases should be within 1–2%, and it is therefore considered negligible for the present study. Much larger differences were recorded by comparing open and closed tunnel measurements and an experimental correction was also proposed in [12]. Such evaluations were instrumental for the present experimental study are recommended to other researchers interested in doing wind tunnel tests of wind turbines.

3.2 The Traversing Systems

The wakes of the turbines were investigated on several surfaces downstream of the rotors. To this end, several pressure and hot-wire probes were traversed on these surfaces, by mounting them on a traversing system that is able to apply the imposed movement. Due to the wide extension of the covered area the design of the traversing system was not negligible and evolved with time from the first to the last test campaign.

Three guidelines drove the design of the traversing system:

- to guarantee the mechanical resistance of the structure for all the positions of the probes;
- to minimize the vibration of the structure, that might affect the quality of the unsteady velocity measurements;
- to minimize the blockage induced by the encumbrance of the traversing.

The very first traversing configuration, manufactured in 2009 to match the aforementioned requirements, was subsequently evolved to on the basis of the experience gained during the tests. Three main configurations were historically manufactured and applied, summarized in Fig. 3.

Figure 3a shows the first layout, used for wake measurements of the H-shape turbine only, subsequently documented in [11]. The traversing allowed the probes to be traversed on a 2D plane, namely in vertical (axial) and transversal (cross-wise) directions, by two computer-controlled actuators; in addition to these movements, another computer-controlled actuator supplied the rotation of one of the probes around its own axis. The probes were mounted on a slender beam that was placed on a support, so to cover the upper half of the wake. When operated, this system demonstrated negligible blockage (even in closed-room tests) and adequate mechanical resistance, but exhibited relatively high vibrations for wind speed exceeding 16 m/s (for the H-shaped turbine rotating at 400 rpm).

Since an interest existed in extending the tests beyond the aforementioned limited, the traversing system was re-designed for the second test campaign (2012, in which

(a): 2009 **(b): 2012** **(c): 2013—2016**

Fig. 3 Pictures of the different traversing systems manufactured for the tests campaigns, during operation

again only the H-shape turbine was studied, and the results documented in [12]); the resulting layout is reported in Fig. 3b. The concept remained the same, with a slender beam bringing the probes sufficiently high to cover the upper half of the wake; however, to limit the vibrations at high wind speed the beam was constrained to move transversally along two parallel runners at the bottom and at the top of the beam. This led to manufacture a 'portal' contouring the vertical-transversal measurement area. This system resulted very effective in reducing the vibrations but originated a significant blockage when the turbine was tested in closed-room environment (especially for near-wake tests, close to the rotor).

With the aim of minimizing both vibration and blockage, a third system was then developed for the tests of 2013 (in which only the troposkien turbine was studied, documented in [13]), later confirmed in the campaign of 2016 (in which both the H-shape and the troposkien turbines were studied). The idea was to return to the original beam system, traversing on a single runner, stiffening the beam and measuring the bottom half of the wakes. Moreover, the probes were mounted on a second beam, so that the vertical beam and the supports were considerably recessed with respect to the probe position and much farther from the rotor. This system resulted stable and with low vibration at high speed (as high as 20 m/s), with negligible blockage. The horizontal beam was equipped with a further actuator to move the probes in horizontal (stream-wise) direction, thus ensuring a 3D traversing of the probes. A second rotating actuator was added to rotate two probes around their own axis, thus leading to a 5 computer-controlled actuators available during the most recent tests.

In this paper, wake measurements performed on straight planes downstream of the rotors are presented. For the H-shape turbine, two measurement surfaces are considered, covering both the top and bottom halves of the rotor, placed 1.5 radii distant from the turbine shaft (i.e., 0.5 radii distant from the cylinder swept by the blades). For the troposkien turbine, only one measurement surface is considered covering the bottom half of the rotor, placed 1.7 equatorial radii distant from the

Fig. 4 Probe system mounted on the traversing in 2016 test campaign

rotor shaft (i.e., 0.7 equatorial radii distant from the circle swept by the blades on the equatorial surface).

3.3 Instrumentation

Pressure, velocity and turbulence measurements were performed using two kinds of probes. A directional *pneumatic five hole probe* measured the time-averaged pressure field and the three-dimensional velocity vector. Two single-sensor *hot wire anemometers* provided the time-resolved stream-wise (i.e., in horizontal direction) and cross-wise (i.e., in transversal direction) velocity components, as well as the intensity of the stream-wise turbulence component. Calibration of the aforementioned probes on a calibrated nozzle lead to estimate an accuracy of 2% in the velocity measurement. Full details on the measurement technique, on the measurement accuracy and on the data reduction are reported in [12, 13]. A picture showing the probes mounted on the traversing system is reported in Fig. 4.

4 Wake Character

The comprehensive nature of this paper and the amount of data available prevents from reporting a complete analysis of the wake measured in all the conditions (that can be found in [11–13]). Conversely, a selection of the most relevant experimental results is shown, with the aim of reviewing the most interesting and general wake features and provide technical guidelines resulting from them.

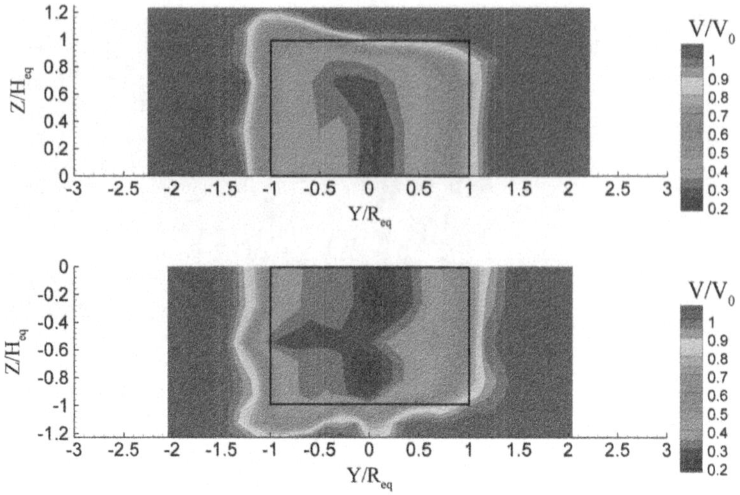

Fig. 5 Wake velocity field for the H-shape turbine at TSR = 2.4. Top: top half of the wake. Bottom: bottom half of the wake

4.1 H-Shaped Rotor

The H-shaped rotor exhibited maximum power coefficient for a tip speed ratio (TSR) of about 2.5, and a significant reduction of performance for lower and higher TSR. Of these two latter conditions, the one at low TSR is the most relevant in urban application, due to the low angular speed experienced by the turbine in the (presumably frequent) start-up/shut down transients. In this work both the optimal and TSR = 1.5 condition are considered.

Figure 5 shows the wake measured for the turbine operating close to the peak Cp condition (TSR = 2.4). The wake is observed to be asymmetric around the turbine axis, due to the asymmetry experienced by the blades in their revolution.

In particular, the left side of the wake (for $Y/R < 0$) results from the section where the blades move windward (where the aerodynamics is more favourable). Conversely, the right side of the wake (for $Y/R > 0$) results from the section where the blades move leeward (where the highest peak values and gradients of incidence and Reynolds number occur). In fact, the asymmetry is limited in the midspan area; in this region, the wake of the shaft is clearly visible all along the span, and it appears slightly shifted towards the right side of the wake. Similar features were observed by other authors [14, 15], thus confirming the general validity of the proposed remarks; it is to be noted that the present measurements provide a full reconstruction of the wake all along the span, for turbines in their real scale configuration and operated at the Reynolds number typical of the microgeneration in the urban environment.

The asymmetry increases away from midspan; at about 50% of the semi-height of the blades the trace of the struts appears on the left side of the wake (visible only in

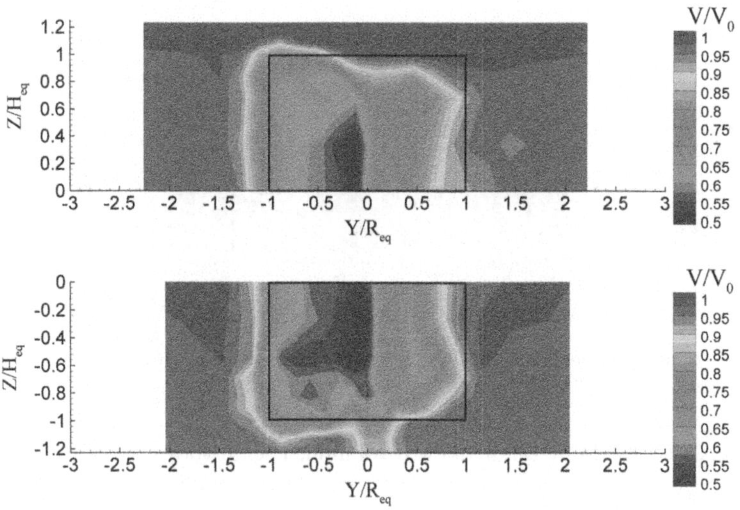

Fig. 6 Wake velocity field for the H-shape turbine at TSR = 1.5. Top: top half of the wake. Bottom: bottom half of the wake

the bottom side of the wake, thanks to higher resolution of the measurement grid on this side of the machine). At the tip, the left side of the wake marks the presence of a wider velocity deficit, which is consistent with the onset of the tip vortex. Flow angle measurements show that the tip vortex, in this area, is clockwise; on the other side of the wake, a much weaker vortex appear, migrated towards midspan and counter-clockwise rotating. The analysis of the blade incidence on the different phases of the rotation suggests that the tip vortices are dominated by the phenomena occurring in the upstream section of the rotation, where the wind velocity and, hence, the aerodynamic forces, are higher.

Figure 6 reports the wake distribution for TSR = 1.5, using the same format of Fig. 5. As well known, for lower TSR the oscillations in blade angle of attack and Reynolds number grow in amplitude, leading to the onset of dynamic stall. In this context, the asymmetry between the blade aerodynamics in the windward and leeward phases grows and, therefore, the asymmetry in the wake is greatly amplified. The wake appears weakened in width and velocity deficit, and it is shifted towards the negative Y direction. As found also for peak Cp condition, the asymmetry further enhances in the tip region, where a much larger velocity deficit appears on the left side of the wake, as a result of the strong tip vortex. The corresponding flow angle distribution, shown in Fig. 7, highlights the presence of the strong tip vortex on the windward side of the wake.

The analysis of the time-resolved velocity signals in the wake (not reported for sake of brevity, the interested reader is referred to [13]) showed a general increase of both periodic and turbulent unsteadiness for low TSR. This effect is connected to the onset of dynamic stall, whose traces appear especially in right side of the wake,

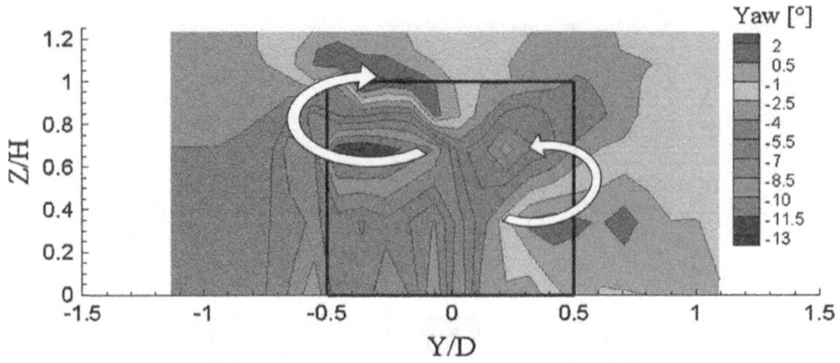

Fig. 7 Flow angle distribution in the top half wake of the H-shape turbine for TSR = 1.5

originated in the leeward phase of the revolution. The tip vortex was also found to be highly unsteady, due to the periodic oscillation of blade loading also in the windward phase of the revolution.

The near wake of the H-shaped turbine was found to roughly follow the shape of the rotor and it still shows the main features of rotor aerodynamics: (i) the asymmetry in working principle between the two main phases of the revolution; (ii) the concentrated losses due to the viscous wakes of the struts and of the shaft; (iii) the asymmetric and highly pulsating the tip vortex, counter-rotating on the two sides of the wake; (iv) the reduction of velocity deficit at low TSR; (v) the increase of unsteadiness for low TSR.

4.2 *Troposkien Rotor*

The wake shed by the troposkien rotor exhibits some similarities to the one of the H-shaped turbine in the equatorial section. Large differences appear close to the turbine tip. Figure 8 reports the bottom half of the wakes measured for peak Cp condition (TSR = 3.1) and low TSR condition (TSR = 2.4).

For both the conditions, the wake follows the highly tapered rotor shape, and it appears shifted towards negative Y direction. The viscous wake of the shaft is clearly visible. As already observed for the H-shaped turbine, a general reduction of the wake velocity deficit occurs as the TSR reduces, especially for $Y > 0$, corresponding (as for the other rotor) to the wake shed during the leeward phase of the revolution movement. The asymmetry between the two wake sides, already visible for peak Cp condition, grows significantly for low TSR.

Moving away from midspan, the character of the wake changes: the velocity deficit reduces and the wake becomes almost symmetric across the shaft. This is due to two main reasons. At first, measuring the wake of a Troposkien turbine on a flat surface leads to increase significantly the distance between the rotor and the probes in the tip regions, and thus the wake measured there has already undergone

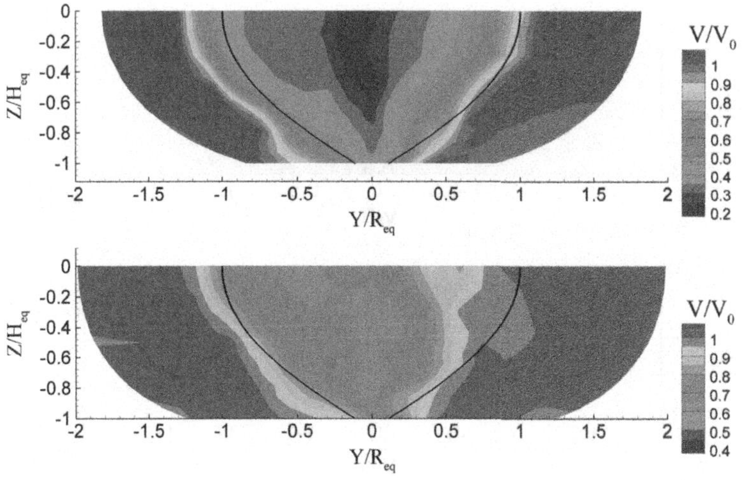

Fig. 8 Wake velocity field for the Troposkien turbine at TSR = 3.1 (maximum Cp, top) and at TSR = 2.4 (low loading, bottom)

a significant mixing process. Secondarily, the local TSR reduces significantly in the tip regions, amplifying the incidence and Reynolds number oscillations during the revolution and making these oscillations more similar between the windward and leeward phases of the rotation. As a further result, the work extraction—and hence the wake deficit—reduces due to the onset of severe dynamic stall on both the side of the rotor, as confirmed by the unsteady velocity components (not reported for sake of brevity, the reader is referred to [13] for full details).

Just above the tip area ($-0.8 < z/H < -0.6$), an accurate inspection of the wake profiles reveals the presence of ripples on the right side margin of the wake, that do not appear on the other side. The distributions of flow angle, reported in Fig. 9 for the two TSR conditions under consideration, indicate the presence of a counter-rotating vortex in this area. Due to the closed-tip configuration of the Troposkien rotor, these vortices cannot be generated by the trailing vorticity as occurred for the tip vortex of the H-shaped turbine. Conversely, these vortices result from the detachment of shed vorticity from the blade profiles, that lie on surfaces non-orthogonal to the turbine axis and, therefore, generate a vorticity component in the stream-wise direction.

It is to be noted, however, that these vortex cores are smaller and weaker than those observed in the wake of the H-shaped turbine and, therefore, are expected to have a lower impact on adjacent objects.

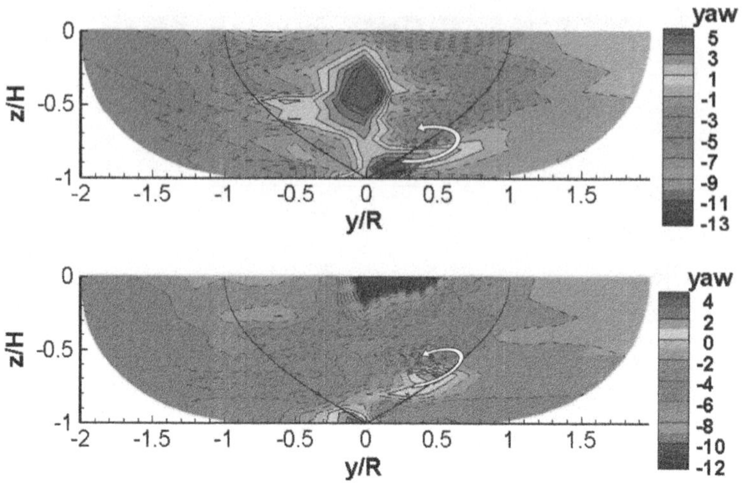

Fig. 9 Flow angle distribution in the wake for the Troposkien turbine at TSR = 3.1 (maximum Cp, top) and at TSR = 2.4 (low loading, bottom)

4.3 *Reynolds Number Effect*

The wake of both the turbines have been observed to change significantly with the TSR. To complete the characterization of the VAWT wakes, it is of interest to also investigate the impact of the Reynolds number on the wake. In practice, the actual Reynolds number at which the blades operated changes during the revolution, but it is not evaluable in the experiments as the actual relative velocity of the flow approaching the blades during the revolution is unknown. As an alternative, a 'peripheral' Reynolds number is usually defined as:

$$Re_U = U_{EQ} D_{EQ} / \nu$$

with U_{EQ} and D_{EQ} the peripheral speed and the diameter on the equatorial section respectively, and ν the cinematic viscosity. In the wind tunnel tests, the TSR was changed by changing the wind speed at constant angular speed (400 rpm), thus keeping constant Re_U. The impact of Reynolds number was, therefore, investigated in a dedicated test campaign in which the angular speed was changed at constant TSR.

The maximum excursion of Reynolds number available in the tests, imposed by the maximum excursion of angular speed available (limited by structural aspects at high speed, and by measurement reliability considerations at low speed), was achieved for the Troposkien rotor, which was operated in the range 250–600 rpm. This resulted in an excursion of Re_U in the range 114,000–273,000 (with 180,000 as reference value of the basic tests at 400 rpm).

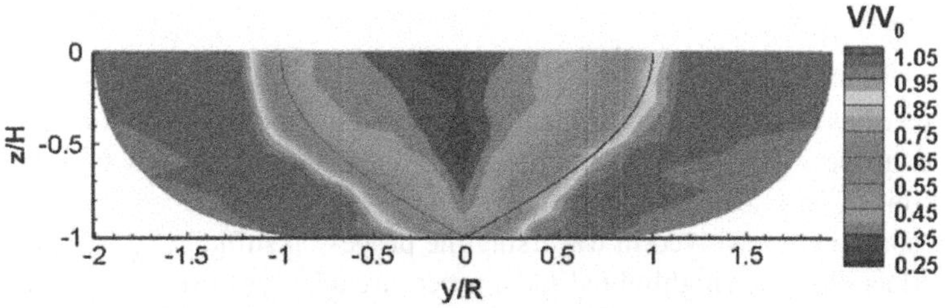

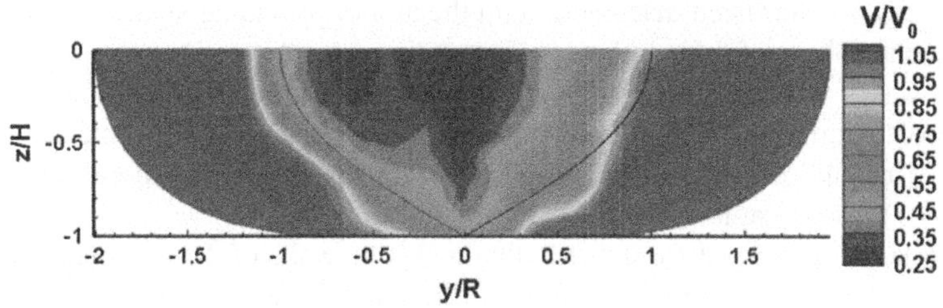

Fig. 10 Impact of Reynolds number on the Troposkien turbine wake. Top: n = 600 rpm, Re_U = 273,000; Bottom; n = 250 rpm, Re_U = 114,000

This range is considered relatively large, especially because of the low general levels of Reynolds number usually found in VAWTs for microgeneration.

Figure 10 shows the wake distributions measured for the Troposkien rotor for the Re_U = 273,000 (top) and Re_U = 114,000 (bottom) for TSR 3.1, which corresponds to peak Cp condition at high Reynolds number while, for low Reynolds number, it is slightly lower to the peak Cp condition (which occur for TSR = 3.3).

In general, the shape of the wake does not change and retains the same features of the one commented before for nominal angular speed, especially for $z/H > -0.5$. However, a significant change in velocity deficit occurs. The velocity deficit is seen to increase significantly with the reduction of Reynolds number, especially in the regions affected by the viscous wake of the shaft—which is, of course, wider and stronger at low Reynolds number. The velocity deficit also increases on the left side of the wake; as performance measurements indicate a lower performance in this condition, it is argued that the increase of velocity deficit is caused by viscous and turbulent effects in the wake stream-wise evolution and not by an enhanced work exchange.

In the tip region of the wake the reduction of Reynolds number has the highest impact, as the wake for Re_U = 114,000 appears more enlarged, probably due to an enhanced mixing process occurring in this area, in which high oscillations occur due to the low local TSR. Unsteady velocity oscillations, not reported for sake of brevity, confirm a significant increase of turbulence unsteadiness on the whole right side of the wake for the lowest Reynolds number.

5 Conclusions

This paper has presented the technical aspects and the most relevant outcomes of a comprehensive wind-tunnel test campaign on the wakes shed by two vertical axis wind turbines for microgeneration.

The complexity involved in traversing the probes on sufficiently large measurement surfaces has been highlighted and the measurement system required by this kind of investigation has been discussed, with the aim of providing significant technical indications to be useful for future similar studies.

The wakes measured downstream of H-shaped and Troposkien turbines exhibited significant similarities. The wakes roughly follow the shape of the rotor, they appear enlarged and almost symmetric for high TSR, while result weaker, more asymmetric, and more unsteady at low TSR.

Despite these general similarities, that make the wake of the two machines very similar in the equatorial zone, significant difference arise between the wakes of the two machines in the tip area. In particular, a large-scale system of tip vortices has been observed in the wake of the H-shaped turbine, as a result of the trailing vorticity detached by the blades; this vortex system changes its sign and its amplitude on the two sides of the wake, resulting much stronger on the side corresponding to the 'windward' blade motion, as a result of the asymmetric working principle of the VAWT. The generation of these vortices is clearly inhibited in the Troposkien turbine, as a result of the tapered shape of the rotor; however, small-scale vortices are observed close to the tip region on the side of the wake corresponding to the leeward phase of the blade rotation. These vortex systems are connected to the shed vorticity, in a region where the tapered shape of the rotor causes a very low local TSR and a three-dimensional vortex detachment process (so that the detached vorticity acquires a stream-wise component).

When operating the Troposkien rotor at very low Reynolds number, the wake properties change significantly, resulting wider, with higher velocity defect and affected by higher turbulent unsteadiness.

On the basis of the observed features, a potential for optimization seems to exist for the placement of multiple turbines in the urban environment. As a matter of fact, the inherent asymmetry of the VAWT wakes might be exploited to reduce the distance between consecutive machines. Moreover, the Troposkien turbine seems particularly interesting as its wake affects a limited amount of space in the tip region, which is also free from wide vortical structures. However, the exploitation of these features requires the availability of reliable prediction tools. From this perspective, the measurements performed on VAWTs at Politecnico di Milano constitute a significant data-base of results. Future works will include the analysis on the stream-wise evolution of the wake on the equatorial surface, also considering time-resolved flow angle measurements and estimates of the turbulent scales.

Acknowledgements The present work is a result of the contributions from the DeepWind project, supported by the European Commission (FP7 Energy 2010 - Future emerging technologies), and the MIUR (Italian Ministry of Education, University and Research). The authors would like to express their gratitude to the Company Tozzi-Nord Wind Turbines, and to the technicians and collaborators of the Politecnico di Milano and the Università di Trento for their support in performing the experiments.

References

1. Balduzzi, F., Bianchini, A., Carnevale, E.A., Ferrari, L., Magnani, S.: Feasibility analysis of a Darrieus vertical-axis wind turbine installation in the rooftop of a building. Appl. Energy **97**, 921–929 (2012)
2. Mertens, S., van Kuik, G., van Bussel, G.: Performance of an H-Darrieus in the skewed flow on a roof. J. Solar Energy Eng. **125**, 433–440 (2003)
3. Mertens, S.: Wind Energy in the Built Environment—Concentrator Effects of Buildings. Multi-science (2006)
4. Lee, K.Y., Tsao, S.H., Tzeng, C.W., Lin, H.J.: Influence of the vertical wind and wind direction on the power output of a small vertical-axis wind turbine installed on the rooftop of a building. Appl. Energy **209**, 383–391 (2018)
5. Battisti, L., Benini, E., Brighenti, A., Raciti Castelli, M., dell'Anna, S., Dossena, V., Persico, G., Schmidt Paulsen, U., Pedersen, T.F.: Wind tunnel testing of the DeepWind demonstrator in design and tilted operating conditions. Energy **111**, 484–497 (2016)
6. Shamsoddin, S., Porté-Agel, F.: Large eddy simulation of vertical axis wind turbine wakes. Energies **7**(2), 890–912 (2014)
7. Mendoza, M., Bachant, P., Wosnik, M., Goude, A.: Validation of an actuator line model coupled to a dynamic stall model for pitching motions characteristic to vertical axis turbines. J. Phys. Conf. Ser. **753**(2) (2016)
8. Schito, P., Zasso, A.: Actuator forces in CFD: RANS and LES modeling in OpenFOAM. J. Phys: Conf. Ser. **524**, 012160 (2014)
9. Maskell, E.C.: A theory of the blockage effects on bluff bodies and stalled wings in enclosed wind tunnel, ARC/R & M-3400 (1963)
10. Mercker, E., Wiedemann, J.: On the correction of the interference effects in open jet wind tunnels, SAE Paper No. 960671 (1996)
11. Battisti, L., Zanne, L., Dell'Anna, S., Dossena, V., Persico, G., Paradiso, B.: Aerodynamic measurements on a vertical axis wind turbine in a large scale wind tunnel. J. Energy Res. Technol. **133**, 031201 (2011)
12. Dossena, V., Persico, G., Paradiso, B., Battisti, L., Dell'Anna, S., Brighenti, A., Benini, E.: An experimental study of the aerodynamics and performance of a vertical axis wind turbine in a confined and unconfined environment. J. Energy Res. Technol. **137**, 051207 (2015)
13. Persico, G., Dossena, V., Paradiso, B., Battisti, L., Brighenti, A., Benini, E.: Time-resolved experimental characterization of wakes shed by h-shaped and troposkien vertical axis wind turbines. ASME J. Energy. Res. Technol. **139**(3), 031203 (11 pages) (2017)
14. Tescione, G., Ragni, D., He, C., Ferreira, C.S., van Bussel, G.: Near wake flow analysis of a vertical axis wind turbine by stereoscopic particle image velocimetry. Renew. Energy **70**, 47–61 (2014)
15. Peng, H., Lam, H., Lee, C.: Investigation into the wake aerodynamics of a five-straight-bladed vertical axis wind turbine by wind tunnel tests. J. Wind Eng. Ind. Aerodyn. **155**, 23–35 (2016)

References

A Review Based on Evaluation Experiences with Ten-Years Activity in VAWT Experimental Wind Tunnel Testing

L. Battisti, E. Benini, A. Brighenti, S. Dell'Anna, M. Raciti Castelli, V. Dossena, G. Persico and B. Paradiso

Abstract The purpose of this paper is to provide guidance on the practice gained on vertical-axis wind turbine (VAWT) testing by a combined group of researchers from the Università di Trento (IT) and the Politecnico di Milano (IT), from early experiences dating back to 2007 up to the present day. The adopted operating procedures are discussed with particular care set to the achievement of high precision measurements to be used both as a benchmark for the validation of numerical codes and as a contribution to a deeper understanding of the flow field around VAWTs, to be converted into novel measurement procedures. As a matter of fact, by providing historical and technical information from the ongoing research activity, the lessons learned about the main obtained results could serve as a valuable tool for use by other research groups who are facing similar activities, providing information on which to help base their project plans.

Keywords VAWT · H-shaped and troposkien rotors · Wind tunnel testing Blockage

1 Introduction and Background

Experimental campaigns on micro vertical axis wind turbines (VAWTs) in controlled conditions started in 2011 at the large scale, high speed wind tunnel of the Politecnico di Milano (IT), focusing on different three bladed Darrieus rotors (an H-shaped rotor [1] and a troposkien one [2], designed and manufactured under the supervision of the Università di Trento, (IT), with a symmetrical profile. A second troposkien three

L. Battisti (✉) · E. Benini · A. Brighenti · S. Dell'Anna · M. Raciti Castelli
Laboratorio interdisciplinare di Tecnologie Energetiche, Dipartimento di Ingegneria Civile, Ambientale e Meccanica, Università di Trento, Via Mesiano 77, 38123 Trento, Italy
e-mail: lorenzo.battisti@unitn.it

V. Dossena · G. Persico · B. Paradiso
Laboratorio di Fluidodinamica delle Macchine, Dipartimento di Energia, Politecnico di Milano, Via Lambruschini 4, 20158 Milano, Italy

L. Battisti and M. Ricci (eds.), *Wind Energy Exploitation in Urban Environment*, Green Energy and Technology, https://doi.org/10.1007/978-3-319-74944-0_10

bladed rotor, with an asymmetrical profile [3], was successively tested as a result of the contributions from the DeepWind project, supported by the European Commission, with the aim of exploring also some crucial technological and operational aspects of floating wind turbines.

The main driver for all such campaigns was the recognition that (i) VAWT behavior and flow field was far to be completely understood; (ii) this was one of the reasons that caused commercial VAWTs fail to become operational [4], in spite of all the research, perseverance and money invested so far. As a matter of fact, a very sensitive point and a recurring matter of discussion was (and still is) the reliability of numerical methods—blade element momentum (BEM), computational fluid dynamics (CFD), vortex models, etc.—in reproducing the main features of the rotor flow field and, consequently, in correctly predicting extreme and fatigue loads for the structural design of the rotor.

Aiming at giving some valuable contribution to a deeper understanding of the complex flow physics of VAWTs, characterized by a continuous variation of each blade angle of attack during its revolution [5] and by a wake meandering downstream with respect to the rotor disc [6], aerodynamic torque and thrust were measured through a high accuracy test bench, while the velocity field in the wake of the tested rotors was fully characterized by means of an instrumented traversing system.

At the initial stage, the investigations concentrated mainly on the effect of blockage, comparing the results obtained for both close chamber and open channel wind tunnel configurations [1, 7], and on the influence of the turbine shape (i.e. straight-bladed versus Troposkien) on its overall performance, also comparing the measured wakes downstream with respect to the rotor test section [2]. The measurements then aimed mostly at obtaining high quality data to be used as benchmark for the validation of numerical codes [3] and, finally, the impact of an inclined flow field on the performance of a troposkien rotor was evaluated, with the purpose of quantifying the effect of a tilted operation for an offshore turbine mounted on a floating platform [3, 8]. Part of the obtained results was also numerically replicated by resorting to CFD analyses [9], providing further information on the flow field.

2 The Experimental Setup and the Most Relevant Measurement Procedures

2.1 *The Wind Tunnel*

The large scale—high speed wind tunnel of the Politecnico di Milano (IT) shown in Fig. 1—consists of a closed-loop facility characterized by a working section of 4.00 m width, 3.84 m height and 6.00 m length. It is capable of providing high wind speeds (up to 55 m/s) by resorting to two arrays of multiple fans working in parallel.

Fig. 1 Schematic view of the large scale wind tunnel of the Politecnico di Milano (IT)

Fig. 2 Side view of the test section with the H-shaped rotor and the traversing system (on the right)

High flow quality is assured by a combination of honeycombs and anti-turbulence screens, leading to very low a freestream turbulence intensity at the rotor test section, less than approximately 1%.

For the here presented campaign of measurements, the wind tunnel was operated both using a classical closed chamber and a "free jet" (open channel) configuration, obtained by removing the test room (upstream the diffuser leading to the arrays of fans) and positioning the rotor directly facing the upstream tunnel. Figure 2 shows a side view of the open test section during the measurements on the H-shaped rotor.

2.2 *The Instrumented Test Bench and the Collection of Aerodynamic Data*

Rotor mechanical loads measurements were performed by resorting to a high accuracy test bench, which was instrumented on a shaft below the lowest blade attachment, using two *strain gauge bridges* (installed on the support mast and capable of detecting both streamwise and transversal aerodynamic thrusts), an *absolute encoder* (to provide both rotational speed and the azimuthal position of a reference blade) and a precision *torque meter* (to measure the mechanical torque at the rotor shaft). In order to monitoring the VAWT structural safety, two axial accelerometers were mounted in the upper part of the supporting mast to check the streamwise and longitudinal vibrations of the structure. Figure 3 shows a schematic description of the instrumented test bench, whose sensors and main characteristics are listed in Table 1.

A synchronous motor/generator was mounted on the power train and controlled by an inverter to operate at a constant rotor angular velocity during each measurement. An aerodynamically-shaped hull was installed around the measurement train, as shown at the bottom of Fig. 3.

The *data acquisition system* provided rough information with a sampling frequency of 2 kHz, converted from voltage or current to derived quantities of interest, such as shaft mechanical torque Q_{mec} and bending moment M_T (provided by strain gauges). The bearing losses Q_{loss} were measured and combined to the mechanical torque and rotational speed ω, computing the aerodynamic power P_{aero} as:

$$P_{aero} = (Q_{mec} + Q_{loss})\,\omega = Q_{aero}\,\omega$$

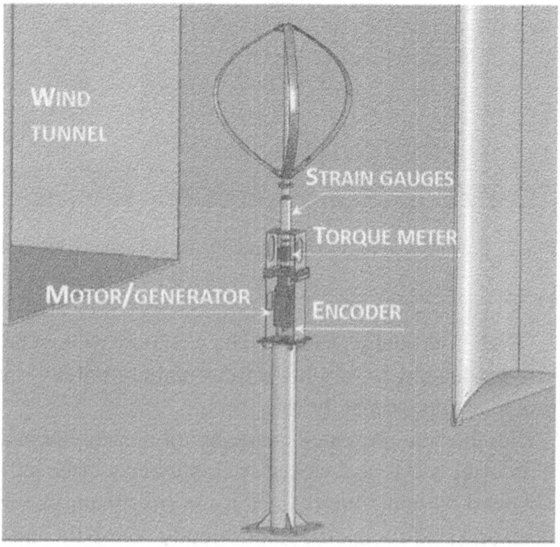

Fig. 3 Scheme of the instrumented test bench

Table 1 Sensors type, measurement range and accuracy

Signal output	Sensor type	Sensor range	Sensor accuracy
Tower bending moment	Strain gauge (full bridge)	±250 Nm	±1.5 Nm
Shaft torque	Torque meter	±200 Nm	±0.1 Nm
Rotational speed	Absolute encoder	Till 6000 rpm	±0.04 rad/s
Blade azimuthal position	Absolute encoder	Full range	±1.8° (at 200 rmp) ±3.6° (at 400 rmp)
Air temperature	RTD PT 100	From −196 to 300 °C	±0.012 °C
Air pressure	Barometer	From 800 to 1060 hPa	±1 hPa

Rotor thrust T was assumed to be acting at rotor midspan (which was located at a certain distance *d* from the strain gauges). It was therefore derived from the bending moment M_T, in formula:

$$T = M_T/d = (A\delta + B)/d$$

being δ the strain gauge full bridge signal, while A and B are the sensor calibration coefficients.

A measurement uncertainty analysis was conducted according to the standard ISO ENV 13005 guide [10]. Even though the methodology is not here reported for brevity's sake, a short discussion is nevertheless proposed to allow a better comprehension of the adopted measurement procedure: for each measurement (characterized by a constant rotor angular velocity and a constant wind tunnel speed), the time series were divided into 1-revolution bins. All bins mean values were then computed, thus leading to a final data being the average of all such mean values (mechanical torque, bending moment, rotational speed and wind speed). The standard deviation and the type A measurement uncertainty were determined. Also the type B uncertainty was evaluated from sensors data sheets. Finally, the combined standard uncertainty of the derived measurements (power, thrust, coefficient of power and thrust) was determined according to Chap. 5 of [10].

2.3 *The Traversing System*

Wake measurements were performed in vertical planes located downstream of the rotor test section by resorting to a motorized sensor arrangement (see Fig. 4) capable of collecting data on both the time averaged and the time resolved flow field. Velocity and turbulence measurements were performed using:

- a directional *pneumatic five hole probe*, for the measurement of the time-averaged pressure field and the three-dimensional velocity vector;

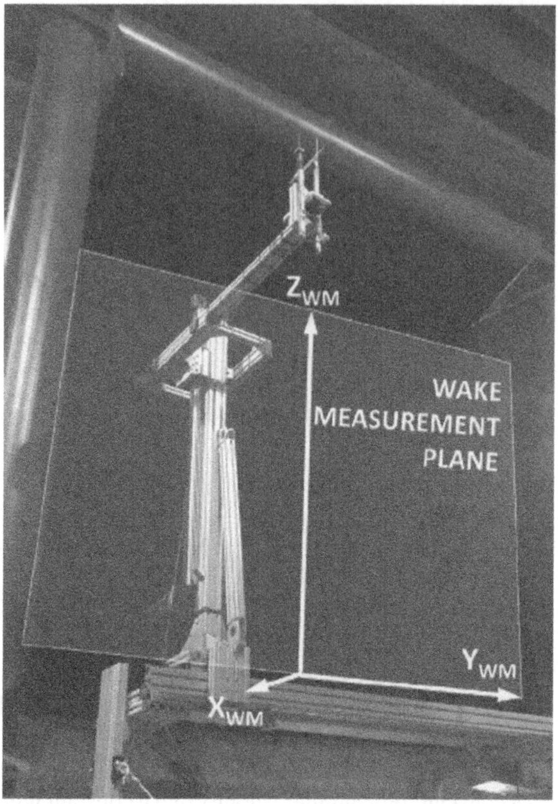

Fig. 4 Motorized sensors arrangement for the measurement of the wakes downstream of the rotors

- two single sensor *hot wire anemometers*, the first probe was mounted in the vertical direction and was rotated in multiple angular positions to achieve, via directional calibration, the time-resolved measurements of velocity magnitude and horizontal flow angle. The second probe was mounted in the streamwise direction, so to achieve a direct velocity measurement avoiding any specific directional calibration.

Such sensors were traversed on a proper measurement grid, extending from the rotor equatorial plane to its blade tip (due to flow field symmetry, only half of the wake was investigated apart from the case of the inclined troposkien rotor, where the measurements extended also in the upper wake portion). For further details, the interested reader is referred to [2, 3].

3 The Tested Rotors

Table 2 summarizes the main geometrical features of the three tested Darrieus rotors. They are all characterized by a three-bladed architecture. The H-shaped rotor and

Table 2 Main geometrical features of the tested rotors

Rotor type	H-shaped	Troposkien shaped N. 1	Troposkien shaped N. 2
Swept area, A (m^2)	1.5	1.5	2.63
Blade profile	NACA 0021	NACA 0021	DU 06-W-200
Blade chord, c (mm)	85	85	101
Blade number, N (–)	3	3	3
Rotor equatorial diameter, D (m)	1.030	1.510	2.028
Rotor height, h (m)	1.457	1.510	1.902
Blade length, L (m)	1.457	2.233	2.945
Rotor solidity, σ (–)	0.25	0.38	0.31

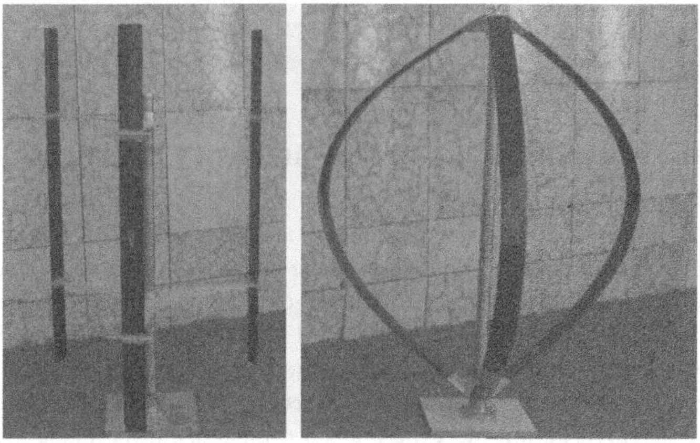

Fig. 5 The H-shaped rotor (left) and the troposkien-shaped rotor (right), both with NACA 0021 profile (N. 1)

the first troposkien one share also the same swept area (1.5 m^2) and blade section (the classical NACA 0021 airfoil characterized by a chord of 85 mm), while the second troposkien rotor is significantly bigger (being characterized by a swept area of 2.63 m^2) and adopts an asymmetrical profile (the DU 06-200).

Rotor solidity is computed according to the equation $\sigma = N\ c\ L/A$. Figure 5 shows a comparison between the H-shaped rotor and the first troposkien one, while the second troposkien rotor is shown in Fig. 6.

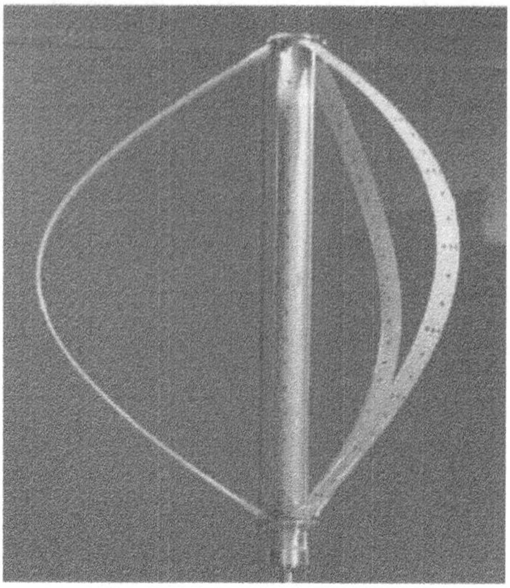

Fig. 6 The troposkien-shaped rotor with DU06W200 profile (N. 2)

4 Main Achievements of the Measurement Campaigns

Although space prohibits a thorough examination of every aspect of all the conducted campaigns of measurements (whose details the interested reader can found in [1–3, 7, 8]), the here presented paper concentrates on a comprehensive review of their major achievements. Even though in outward appearance this does not look very different from a mere list of the obtained results, the authors aim to collect a certain number of very specialized information to document the stages of evolution of the knowledge acquired during the experimental campaigns, to be made available to other research groups who are facing similar activities, providing information on which to help base their future project plans.

4.1 Review of Blockage Ratios Adopted During Experimental Tests from Literature

Since the wind tunnel test section is characterized by a confined volume, aerodynamic measurements could not resemble those obtained during open field operation and researchers usually have to deal with the necessary correction of the wind tunnel blockage. From a preliminary literature review, it clearly appeared that too high geometric blockage factors were often used (see e.g. [11, 12]), thus leading to potentially altered power curves with respect to those resembling open field operation.

In order to assess the behavior of the tested rotors in an open-field environment, a preliminary quantification of blockage correction in open-tunnel can be performed applying the Merker and Wiedemann model [3, 13]. As collected in Table 3, the

Table 3 Blockage of the tested rotors

Rotor type	H-shaped (%)	Troposkien shaped N. 1 (%)	Troposkien shaped N. 2
Solid blockage, $A/A_{chamber}$ (–)	9.7	9.7	17.0
Blockage correction, V'_0/V_0 (–)	1.5	2.1	3.2

blockage correction, defined as the ratio between the open-field wind speed (V'_0) and the open-jet wind speed (V_0) is sufficiently small to avoid any correction on the measured quantities.

4.2 Critical Analysis of Blockage Models from Literature

Aiming at deeply investigating the effect of blockage on the aerodynamics of the tested rotors, the H-Darrieus was operated in both closed and open (free jet) wind tunnel configurations, in such a way to quantify the blockage for different operational conditions [1, 7].

The *blockage correction coefficient* is defined as the ratio between the velocity, V_0', measured in the open jet wind tunnel and the one measured using the closed chamber wind tunnel configuration, V_0, which determine the same thrust on the rotor. Literature correlations usually adopted to account for blockage effects are the original Glauert correlation [14], and the one proposed by Mikkelsen and Sørensen [15, 16].

Figure 7 shows the evolution of the experimental and theoretical blockage correction coefficients as a function of the thrust coefficient, C_T. It can be clearly observed that the adopted correction models do not provide accurate estimates of the blockage effect: as a matter of fact, both the Glauert model and the Mikkelsen and Sørensen model correctly predict the increase of blockage at lower thrusts (<0.7). This suggests that, as the angular velocity increases with respect to the wind speed, the rotor becomes more "solid" and, as a consequence, the amount of flow that transits within the virtual cylinder traced by the blade motion reduces and a larger amount of flow is forced to move around the rotor itself. On the other hand, both models dramatically fail in reproducing the shape of the whole curve.

Such behavior can be explained by considering that the complex fluid-dynamics of a VAWT is barely modeled by a simple 1D momentum approach. The relevant discrepancies registered between measured data and the available theoretical models suggest that specific corrections should be developed for VAWTs, based on the development of an "actuator cylinder" approach, which appears as much more suitable (due to the three-dimensionality of both the rotor and its wake) than classical

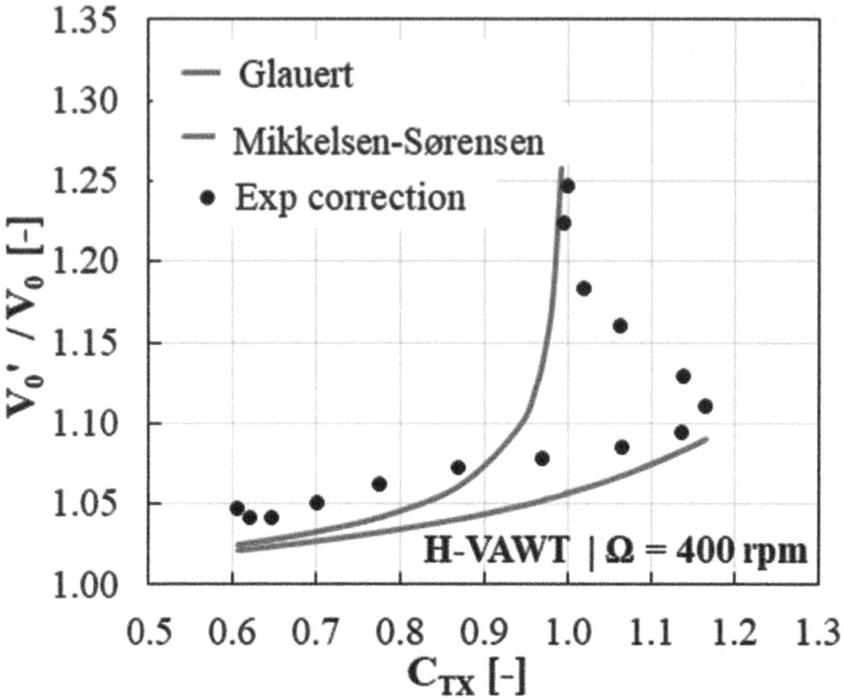

Fig. 7 Wind speed correction factors as a function of the thrust coefficient; comparison between measured values and the most adopted theoretical correction models from literature for the H-shaped rotor

modeling tools based on a simple actuator disc (for more details, the reader is referred to [1, 7]).

4.3 *Performance and Wake of the H-shaped and Troposkien Rotors*

The purpose of the here described campaign of measurements was to shed some light on the main differences between two of the most adopted VAWT architectures. To accomplish this, an H-shaped rotor and a troposkien one, sharing the same swept area, as well as blade number (N = 3) and blade section (NACA 0021, chord 85 mm), were tested using the open jet wind tunnel (to further reduce the effect of blockage).

Performance measurements (see Fig. 8) clearly showed that the troposkien rotor outperforms the H-shaped one (in terms of power exchange), mostly thanks to its larger midspan section (resulting by the constraint imposed on the swept area and

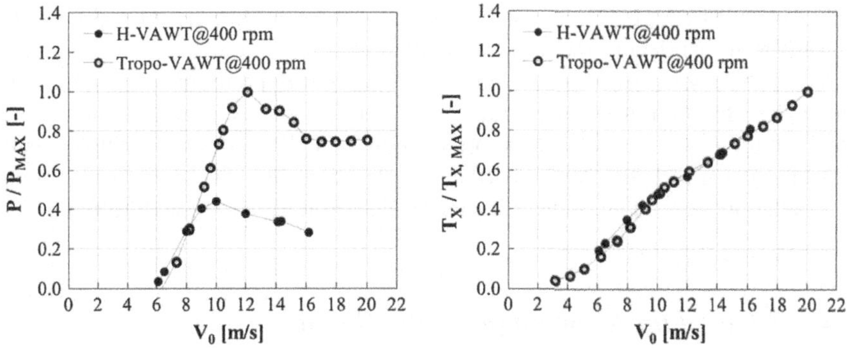

Fig. 8 Measured aerodynamic power and thrust for the H-shaped rotor and the troposkien one as a function of the wind speed for a constant angular velocity of 400 rpm (the maximum refers to the troposkien rotor)

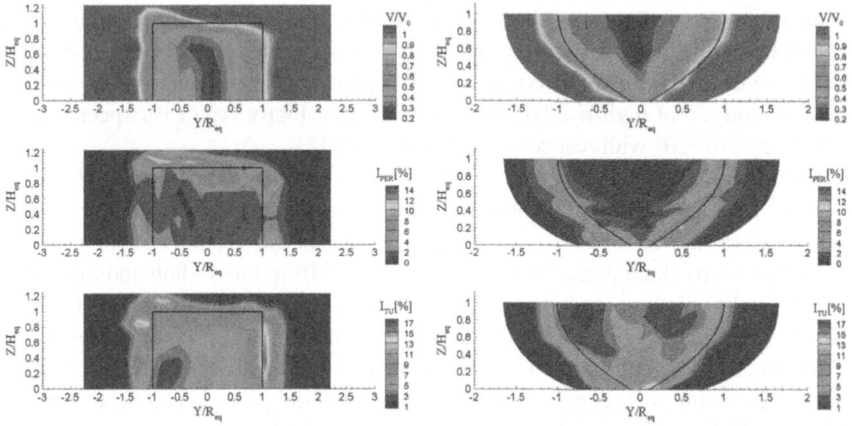

Fig. 9 Wake comparison between H-shaped (left) and troposkien (right) rotors at optimum tip speed ratio, where H is the half-blade height, R is the maximum rotor radius, I_{PER} is the root mean square of periodic unsteadiness, and the I_{TU} is the turbulence intensity [2]

determining a higher Reynolds number) and to the absence of non-aerodynamic struts (which appear to affect the H-shaped rotor with a significant loss penalty) [2].

The above described features resulted consistent with the character of the wakes shed by the rotors, as shown in Fig. 9. In particular, large scale vortices (pulsating significantly during each revolution, due to the periodic fluctuation of the blades aerodynamic loading) dominated the tip region of the wake shed by the H-shaped rotor.

Conversely, the highly tapered shape of the troposkien rotor prevented the onset of tip vortices; on the other hand, the dramatic reduction of tip speed ratio passing from the equatorial section to the tip ones promoted the onset of local dynamic stall, characterized by high periodic and turbulent unsteadiness in the tip region of the wake (for further details, the reader is referred to [2]).

All of the above taken into account, the higher power performance, combined to the higher structural resistance with respect to the H-shaped machine, could make the Troposkien architecture a favorable solution for VAWTs, despite the higher manufacturing cost and complexity. Further investigation should be performed, aiming at a better understanding of the behavior of such promising rotor architecture.

4.4 Effect of the Skew Angle on the Operation of a Troposkien Rotor

Even though the low center of gravity that characterizes VAWTs is generally capable of generating a large overdriving moment, floating troposkien rotors for offshore installations are expected to tilt during operation, acting as a gyro and describing an elliptical trajectory on the water plane. As far as the dynamics of the interaction between wind and rotor is concerned, a quite similar situation is registered for VAWTs operating on top of high-rise buildings, where the velocity vector is expected to be skewed by 10°–20° with respect to the horizontal [17].

Such issues raise several concerns about the aerodynamic behavior of the rotor, prompting for further investigation aiming at a deeper comprehension of the operation of VAWTs in skewed flows. Figure 10 shows a view of the tested troposkien VAWT, characterized by three glass fiber blades, extending from hub to hub and made of a DU 06-W-200 blade section. It was mounted on a high precision test bench, whose axis was suitable to be inclined up to 15° with respect to the upright configuration.

Measured data, collected for a 15° inclination of the shaft with respect to the vertical direction, were consistent with the simple model proposed by Scheurich and Brown [18] for evaluating the performance of a VAWT in oblique flow (based on the swept wing theory, see [19, 20]). According to such theory, the component of wind speed parallel to the shaft has no effect on rotor aerodynamics, thus determining a reduction of the effective unperturbed velocity perceived by the turbine (and, consequently, an increase of the effective tip speed ratio).

As can be inferred from Fig. 11, the overall aerodynamic performance of the rotor operating in skewed flow is lowered by the combined effects of reduced effective velocity and increased effective tip speed ratio. Nevertheless, if the reduction dependent on the perceived wind speed always decreases the aerodynamic efficiency over the whole range of operating conditions, the higher effective tip speed ratio could also increase the rotor performance (depending on the sensitivity of the rotor power coefficient on its tip speed ratio). Interestingly, the model from Scheurich and Brown [18] was able to predict the decay of rotor performance due to tilting in relatively good approximation, as the theoretical results lie within the band of uncertainty of tilted rotor measurements for most of the operating range. This was probably dependent on the troposkien architecture, which minimizes the sensitivity of rotor aerodynamics on skew; in such a configuration, it is reasonable that the main impact of tilting affects the effective incoming velocity. As the geometry of the troposkien rotor is roughly comparable to a sphere, a very small variation of the energy extraction area was registered as a consequence of the shaft inclination. Hence, contrarily to what found by authors focused on H-VAWTs [17], no appreciable increment of power output was obtained when passing from upright to tilted operation.

5 Concluding Remarks: What Lessons Can Be Learned from the so Far Performed Campaigns of Experimental Tests?

The present paper presents the experimental procedure and some experimental results obtained from wide wind tunnel experimental investigations on full-scale VAWTs for microgeneration.

The here discussed project certainly helped improving the understanding of the flow physics associated with VAWT operation. This mainly derived from the fact that the campaigns were not limited to simple performance measurements (power versus wind speed curves), but deployed a sensors net able to measure also the thrust acting

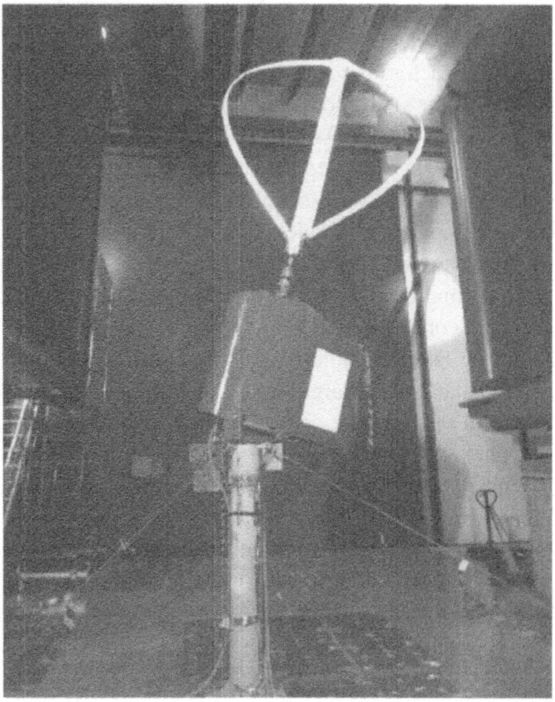

Fig. 10 Side view of the tilted rotor arrangement inside the wind tunnel open chamber

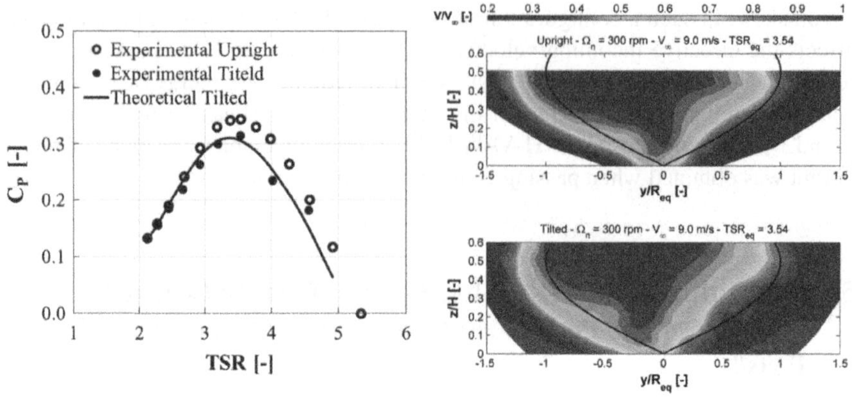

Fig. 11 (right) Measured power coefficient as a function of the tip speed ratio for both upright and 15° tilted rotor operation at 300 rpm [3]; the power coefficient computed according to the simple model from Scheurich and Brown is also shown; (left) time-mean velocity (normalized with the unperturbed wind speed at the test section entrance) on a vertical plane located downstream of the test section, for different operating conditions (the projection of rotor swept area is also represented by means of continuous black lines)

on the rotors and the flow field in the wake. It clearly appears that the absence of such a comprehensive approach would have determined an important missed opportunity.

The aerodynamic axial thrust was measured in both a confined and an unconfined environment on the H-shaped turbine, in order to compute the **experimental blockage** factor between closed and open chamber configurations. Results show that that the rotor solid blockage (almost 10%) introduces a non-negligible interference with the wind tunnel walls. Moreover, the experimental blockage correction factor is found to strongly depend on the operative condition and rotor shape. Lastly, the comparison with well-established blockage models indicates the need of specific blockage correction models for VAWTs.

Performance and wake measurements show that the rotor shape and the operational conditions had a key impact on the turbine performance and on the wake features, contrarily to what happens to thrust loads.

At the same rotational rotor speed, the troposkien rotor outperforms the H-shaped turbine thanks to its larger midspan section (that produces a larger Reynolds number) and to absence of the additional drag contributions (to the non-aerodynamic struts and blade-strut attachments) that characterize the H-shaped rotor. The unsteady and asymmetric work exchange typical of all VAWTs are clearly evidenced by the wake measurements.

Besides the here presented initial findings, the present experimental data-base allows to carry out a deeper investigation on the rotor and wake behavior to serve as to experimental benchmark, in order to complement and support validation and calibration of numerical codes.

Acknowledgements The present work is a result of the contributions from the DeepWind project, supported by the European Commission (FP7 Energy 2010—Future emerging technologies), and the MIUR (Italian Ministry of Education, University and Research). The authors would express their gratitude to the Company Tozzi-Nord Wind Turbines, the technicians and collaborators of Politecnico di Milano and Università di Trento for their support in performing the measurement campaigns.

References

1. Battisti, L., Zanne, L., Dell'Anna, S., Dossena, V., Persico, G., Paradiso, B.: Aerodynamic measurements on a vertical axis wind turbine in a large scale wind tunnel. J. Energy Res. Technol. **133**, 031201 (2011)
2. Persico, G., Dossena, V., Paradiso, B., Battisti, L., Brighenti, A., Benini, E.: Time-resolved experimental characterization of the wakes shed by H-shaped and troposkien vertical axis wind turbines. J. Energy Resour. Technol. **139**(3) (2017). https://doi.org/10.1115/1.4035907
3. Battisti, L., Benini, E., Brighenti, A., Raciti Castelli, M., Dell'Anna, S., Dossena, V., Persico, G., Schmidt Paulsen, U., Pedersen, T.F.: Wind tunnel testing of the DeepWind demonstrator in design and tilted operating conditions. Energy **111**, 484–497 (2016)
4. Sutherland, H.J., Berg, D.E., Ashwill, T.D.: A Retrospective of VAWT Technology. In: SAND2012-0304, Jan 2012
5. Ferreira, C.J.S., van Bussel, G.J.W., Scarano, F.: 2D PIV visualization of dynamic stall on a vertical axis wind turbine. In: 45th AIAA Aerospace Sciences Meeting, Reno, Nevada (US), 8–11 Jan 2007
6. Trivellato, F., Raciti Castelli, M.: Appraisal of Strouhal number in wind turbine engineering. Renew. Sustain. Energy Rev. **49**, 795–804 (2015)
7. Dossena, V., Persico, G., Paradiso, B., Battisti, L., Dell'Anna, S., Brighenti, A., Benini, E.: An experimental study of the aerodynamics and performance of a vertical axis wind turbine in a confined and unconfined environment. J. Energy Res. Technol. **137**, 051207 (2015)
8. Battisti, L., Benini, E., Brighenti, A., Raciti Castelli, M., Dell'Anna, S., Dossena, V., Persico, G., Schmidt Paulsen, U., Pedersen, T.F.: Normalized performance and load data for the deepwind demonstrator in controlled conditions. Data in Brief **8**, 1120–1126 (2016)
9. Raciti Castelli, M., Masi, M., Battisti, L., Benini, E., Brighenti, A., Dossena, V., Persico, G.: Reliability of numerical wind tunnels for VAWT simulation. J. Phys: Conf. Ser. **753**, 082025 (2016)
10. International Organization for Standardization: ISO ENV 13005. Guide to the Expression of Uncertainty in Measurements, technical report, May 1999
11. Van Bussel, G.J.W., Mertens, S., Polinder, H., Sidler, H.F.A.: The development of turby, a small VAWT for the built environment. In: Proceedings of the Global Wind Power Conference and Exhibition, Chicago (US), 28–31 Mar 2004
12. Li, Y., Tagawa, K., Liu, W.: Performance effects of attachment on blade on a straight-bladed vertical axis wind turbine. Curr. Appl. Phys. **10**, S335–S338 (2010)
13. Mercker, E., Wiedemann, J.: On the Correction of the Interference Effects in Open Jet Wind Tunnels, SAE Paper No. 960671 (1996)
14. Glauert, H.: The Elements of Aerofoil and Airscrew Theory, 2nd edn. Cambridge University, Cambridge (GB) (1947)
15. Mikkelsen, R., Sørensen, J.N.: Modelling of wind tunnel blockage. In: Proceedings of the Global Windpower Conference and Exhibition (2002)
16. Sørensen, J.N., Shen, W.Z., and Mikkelsen, R.: Wall Correction Model for Wind Tunnels With Open Test Section. AIAA J. **44**(8), 1890–1894 (2006)
17. Mertens, S., van Kuik, G., Van Bussel, G.: Performance of an h-darrieus in the skewed flow on a roof. J. Solar Eng. 433–440 (2003)

18. Scheurich, F., Brown, R.: Vertical-axis wind turbines in oblique flow: sensitivity to rotor geometry. In: EWEA 2011 Conference, Brussels (BE)
19. Betz, A.: Applied airfoil theory. Unsymmetrical and non-steady types of motion. Aerodynamic Theory IV, 1935, pp. 97–107
20. Jones, R.: Wing Theory. Princeton University Press (1990)

Small Wind Turbine Performance Assessment for Canada

A. Medd and D. H. Wood

Abstract This paper investigates the performance of small wind turbines across Canada through an analysis of wind resource and turbine performance data. Four sets of data from 25 installations have been analyzed: 12 data sets from a Wind Energy Institute of Canada (WEICan) small wind project completed in Western Canada, 5 from small wind test projects at the WEICan test site on Prince Edward Island; and 8 collected sporadically across Prince Edward Island. The results are compared to previous studies in the U.S.A., Ireland, and England. The consistent finding is that some turbines performed well but there is a pattern of low availability, poor power performance, and low capacity factor. Some possible reasons for the poor performance are advanced.

Keywords Small wind turbines · Power output · Field testing
Availability · Power curve · Capacity factor

1 Introduction

While the large wind turbine industry is booming in North America and elsewhere, the small wind turbine industry is not. There was an installed distributed wind capacity of 28 MW in the U.S.A. in 2015 but only 4.3 MW of small wind capacity, defined as turbines with rated power of 100 kW or less, Orrell and Foster [1]. This limit on the size of small wind turbines includes a locally-made turbine but exceeds the

A. Medd
Wind Energy Institute of Canada, Tignish, Canada

Present Address
A. Medd
WSP, Calgary, Canada

D. H. Wood (✉)
Department of Mechanical and Manufacturing Engineering, University of Calgary, Calgary, Canada
e-mail: dhwood@ucalgary.ca

L. Battisti and M. Ricci (eds.), *Wind Energy Exploitation in Urban Environment*,
Green Energy and Technology, https://doi.org/10.1007/978-3-319-74944-0_11

IEC Standard 61400-2 [2] limit of a rotor swept area of less than 200 m^2, or a rated power less than approximately 50 kW. For convenience we will continue to use the larger upper limit. The small wind industry began to contract in 2013, with investment dropping to $21 million in 2015 from $36 million in 2013, Orrell and Foster [1]. This was dramatically illustrated by the sudden closure of the market leader Southwest Windpower in 2013[1] and the leading Canadian manufacturer, Endurance, went bankrupt in 2016.[2] There is no similar data for small wind and distributed capacity and the industry for Canada but the available evidence suggests that some of the underlying reasons, such as poor performance and low availability, are similar to those in the U.S.A. The purpose of the current research is to attempt to quantify the operational problems with small wind turbines and seek explanation of them.

A strong case can be made that small wind is failing partly because of the increasing competitiveness of distributed solar power. California for example, currently has over 660,000 small scale photovoltaic (PV) installations.[3] PV offers high reliability, very low maintenance, relatively predictable power output, no noise issues, and less visual impact. Also important is the dramatic fall in PV prices, by a factor of over two since 2009, Fu et al. [3], which has not been matched by small wind. The competition faced by small wind is intense; significant improvements in cost-effectiveness, performance and reliability are required to meet these challenges.

Several studies have investigated the field performance of small wind turbines. Twenty-one free-standing turbines (i.e. not building mounted) of 10 kW rated power or less in Massachusetts were found to be producing approximately 29% of their predicted energy, Shaw et al. [4]. No wind speed measurements were made at the individual turbines, so considerable uncertainty must be attached to this result but it is noteworthy that no turbine performed better than anticipated, as would likely occur if the wind speed errors were random. This suggests generic problems with the turbines. Reference [4] attributed this under-performance to a combination of factors, including wind map inaccuracy in determining the wind resource, annual variations in wind speed, inaccurate turbine performance data including power curves, and inaccurate site modeling assumptions (e.g., wind shear, obstacles). The Warwick trials in the United Kingdom, Encraft [5] and James et al. [6], also observed under-performance of building mounted turbines by more than 50% in many cases. Stand-alone small turbines in rural locations generally performed better, Sissons, [7]. Pagnini et al. [8] monitored several free-standing turbines in urban locations and found that turbulence appeared to significantly influence the power output. It is clear that the wind resource, particularly in urban areas, is more difficult to estimate accurately and cheaply than the solar resource. For example, the clean energy software tool RETscreen[4] provides monthly averaged solar insolation and 10 m wind speeds for Calgary at one location

[1]See, for example: http://www.irecusa.org/2013/05/what-happened-to-southwest-wind-power/ accessed 3 May 2017.

[2]See, for example: http://vancouversun.com/business/local-business/award-winning-surrey-supplier-of-global-wind-turbines-goes-bankrupt accessed 3 May 2017.

[3]See http://www.gosolarcalifornia.ca.gov/ accessed 3 May 2017.

[4]http://www.nrcan.gc.ca/energy/software-tools/7465 accessed 4 May 2017.

only, the airport. The authors obtained over 20 years of high quality wind and solar data from the University of Calgary weather station, approximately 10 km distant from the airport. The two sites had the same monthly average insolation to within 1%, but vastly different wind speeds. Because average wind speed has a significant effect on average power output as will be documented in Sect. 4, it is important to determine it accurately but cheaply. It is likely, however, that poor quality wind resource assessment will lead to as many under-predictions as over-predictions but none of the field trials we know of have demonstrated over-performance of the turbine relative to the manufacturer's power curve combined with the initial resource assessment. We conclude that factors related to turbine design and technology are the main causes of under-performance.

This paper is organized as follows. The next section describes the field data. This is followed by a description of the analysis methods to determine capacity factor, CF, and availability. The results are presented in Sect. 4 and the conclusions in the last section.

2 Field Data for This Study

The new data presented here comes from: the study of 12 turbines from the Wind Energy Institute of Canada (WEICan) unpublished small wind project completed in Western Canada; 5 time series datasets from WEICan small wind test projects; and 8 multi point datasets collected sporadically across Prince Edward Island (PEI) from 50 kW turbines. For comparison 16 datasets from the Irish Wind Trials.[5]

For the Western Canada study, 1 min wind speed and direction was collected at a height as close as possible to hub height. NRG 40C anemometers and NRG 200P direction vanes were typically used with Campbell Scientific CR1000 loggers. Met towers were within 2–4 rotor diameters of the tower as required by the IEC standard for power performance measurement, [9].

Data collection was typically for 1–6 months only. This should not influence the calculated power curve but would influence any calculation of annual energy production, AEP. For this reason, AEP is not shown here. IEC [9] states that data be excluded when:

- The turbine was not producing due to low wind speeds.
- External conditions other than wind speed are out of the operating range of the wind turbine (e.g. due to fog).
- The turbine had a fault.
- The turbine was manually shut down or in a test of maintenance operating mode.

[5]The performance data for the Irish trials were made available informally to WEICan. Subsequently we have not found any published references for the data and our e-mail enquiries to the Sustainable Energy Authority of Ireland and the consultant who undertook the study and drafted the final report have not been answered. The data appear to be of high quality and consistent with the Canada results. None of our conclusions would be altered if the Irish data were not considered.

Table 1 Western Canada capacity factor and availability

Turbine	1	2	3	4	5	6	7	8	9	10	11	12
Capacity factor	0.08	0.05	0.00	0.25	0.18	0.02	0.02	0.00	0.06	0.04	0.01	0.09
Availability	0.98	0.96	0.11	0.77	0.84	0.22	0.53	0.01	0.91	0.51	0.06	0.94

- Failure or degradation of test equipment (e.g. due to icing).
- The wind direction was outside the "disturbed" measurement sector(s) as defined in [9]. The disturbance is due to obstacles causing unusually high turbulence and possibly low mean wind speeds.

Manual data quality control was also used to filter corrupt data. Any significant and unexplained fluctuations in data were omitted.

The measured and manufacturer's power curves were non-dimensionalized to facilitate the comparison of turbines of different power ratings and to maintain anonymity among wind turbine manufacturers.

3 Analysis

Because there is a wide range of problems that can affect turbine performance the analysis was kept simple.

33 sets of 1 min data for turbines ranging in capacity from 1.3 to 50 kW were analyzed. The data is from the Western Canada project, summarized in Table 1, the Irish Wind Trials (Tables 2a, 2b), select WEICan certification data (Table 3), and a number of 50 kW turbines installed on PEI, (Table 4). CF and availability was calculated for every data set. The former is simply the average power output divided by the rated power and is a measure of the site wind resource as much as the turbine performance. For comparison, the CF dependence on wind speed of the Skystream 2.4 kW wind turbine determined from the manufacturer's power curve is given in Table 5. The analysis is described in Chap. 12 of Wood [10]. The nearly linear dependence of CF on average wind speed demonstrates the point made about their relationship in a footnote on a previous page. It is a general rule-of-thumb for small wind turbines that a "good" hub height average wind speed is 5 m/s which Table 5 shows gives a CF ~ 0.20 which we will take as indicative of "good" performance.

Table 2a Irish wind trials capacity factor and availability

Turbine	1	2	3	4	5	6	7	8
Capacity factor	0.18	0.12	0.09	0.17	0.15	0.22	0.07	0.09
Availability	0.97	0.51	0.98	1.00	0.97	1.00	0.57	0.78

Table 2b Irish wind trials (completed) capacity factor and availability

Site index	9	10	11	12	13	14	15	16
Capacity factor	0.07	0.19	0.26	0.29	0.22	0.20	0.15	0.08
Availability	1.00	0.96	0.84	0.99	0.83	0.89	0.97	0.47

Table 3 WEICan test site (PEI) capacity factor and availability

Turbine	1	2	3	4	5
Capacity factor	0.25	0.07	0.27	0.05	0.25
Availability	0.84	0.55	0.50	0.31	0.61

Table 4 PEI 50 kW Turbine Capacity Factor and Availability

Turbine	1	2	3	4	5	6	7	8
Capacity factor	0.07	0.07	0.08	0.17	0.24	0.16	0.37	0.07
Availability	0.58	0.62	0.62	0.82	0.88	0.56	1.00	0.52

Table 5 Capacity factor dependence on wind speed for the Skystream 2.4 kW turbine

Wind speed (m/s)	2	3	4	5	6	7	8	9	10
Capacity factor	0.01	0.035	0.092	0.177	0.259	0.327	0.386	0.429	0.457

Availability was estimated as the fraction of data points where the power exceeded 1 W while the wind speed was above cut in, typically assumed to be 2 m/s higher than the value quoted by manufacturers as it is well known that small wind turbines start at wind speeds higher than the cut-in speed determined from the power curve, e.g. [10]. This method was used because there is no other way to determine from the data provided when the turbine was broken down and unavailable. Availability gives additional information to CF, because it is possible for a turbine to underperform relative to the manufacturer's power curve but be mechanically robust and so have a relatively high availability.

4 Results

The initial data analysis suggested no correlation between availability or CF with turbine rated power but, for brevity, the results are not shown. Figure 1 plots CF against availability. The average availability from Tables 1, 2a, 2b, 3 and 4 is 0.7 and average CF is 0.13. The data from the Irish wind trials appears to indicate better overall performance. This may be due to better maintenance as well as generally higher wind speeds. In addition a number of the turbines monitored in the Irish wind trials exceeded the manufacturer's power curve, at least for a considerable range of wind speeds. If the data from the Irish wind trials is ignored, the average availability

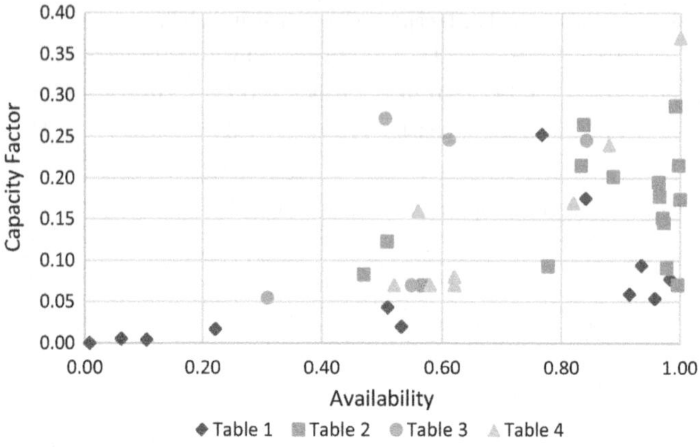

Fig. 1 Availability versus capacity factor from Tables 1, 2a, 2b, 3 and 4

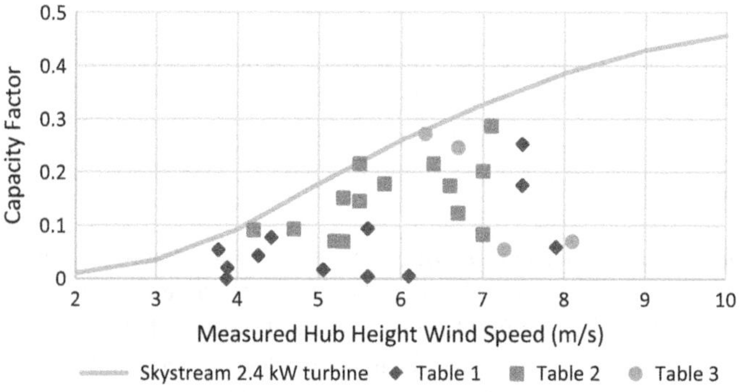

Fig. 2 Capacity versus wind speed factor from Tables 1, 2a, 2b, 3 and 4

drops to 0.61 and the average capacity factor falls to 0.12. From the data, there is a correlation between availability and capacity factor, as expected.

Figure 2 shows CF against mean wind speed for the Skystream 2.4 kW turbine as provided by the manufacturer and the measured CFs for the turbines in this study. The Skystream CF shows the strong dependence of average output on wind speed that was mentioned previously. This not a cubic dependence as sometimes claimed, but a linear one with high slope. For example increasing the wind speed from 5 to 6 m/s gives nearly a 50% increase in average power. It is clear from Fig. 2 that there is a significant problem with small wind turbines that is not due to availability as a number of small wind turbines from different manufacturers with different rated capacities were not performing to expectation. There are a multitude of reasons for these problems, however, it is difficult to diagnose without completing more research on each individual turbine.

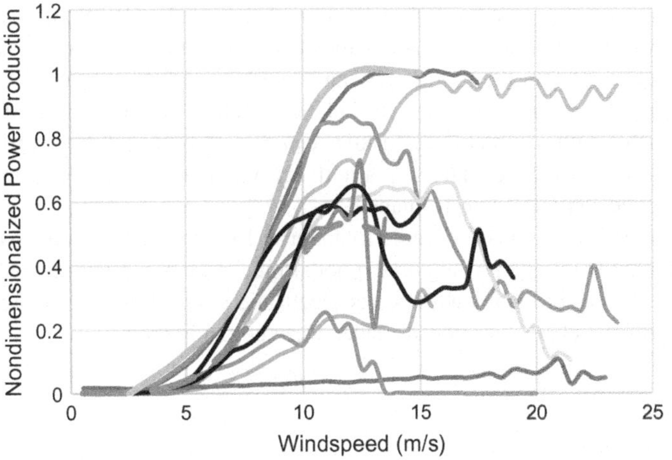

Fig. 3 Power curves from Table 1. The red line is the manufacturer's curve for the Skystream 2.4 kW turbine and the blue dashed line is the average of the Western Canada results. *Note* that all power curves have been normalized to a maximum value of unity

Ten Western Canadian sites provided IEC-compliant power curves that are shown in Fig. 3. The data was non-dimensionalized to the rated capacity of the turbine to ensure that manufacturers are not identified. An average of all power curves is also shown in Fig. 3. A moving average was used to smooth the average curve. Some turbines performed but a significant number produced significantly less power than claimed by the manufacturer.

The Western Canada sites also showed a deviation in CF when compared to CF calculated with WindPro[6] or RETScreen. On average, WindPro overestimated CF by 50% and RETScreen overestimated AEP by 56%. The estimate inaccuracy could be due to any number of factors including poor turbine availability, low quality meteorological data input to the models and inaccurate power curves depending on the site in question.

Three sites from the Western Canada data will be investigated further in this section. All three sites had availabilities over 0.9 and capacity factors less than 0.1 and had detailed site information including pictures and turbulence data in some cases. In two cases the wind turbines were surrounded by low brush with buildings approximately 100 m away. The third turbine was immediately adjacent to a hedge row with buildings nearby, this turbine was exposed to a low roughness field to the west. These installations are similar to those studied by Sissons et al. [7].

In all three cases, the low capacity factors of 0.09, 0.15 and 0.17 are attributed to siting, the turbulence caused by nearby obstacles and high surface roughness coupled with typically low tower heights associated with small wind turbines. In all three sites

[6]http://www.emd.dk/windpro/

the measured average wind speed was within 1 m/s of the Canadian Wind Atlas[7]. WindPro was used to interpolate the values from the Canadian Wind Atlas at 50 m to hub height based on the topography and roughness of the turbine site. The two sites that had turbulence data show turbulence intensities in the range of 15–20%. Two other sites with acceptable performance and availability exhibited turbulence in the range of 10–15% which agrees qualitatively with the finding of Pagnini et al. [8] that turbulence can reduce performance.

In summary, the main contributors to underperformance were found to be: deviation from manufacturer's power curve, poor siting, and low availability. Other possible factors such as problems with power electronic issues, low generator efficiency, and aerodynamic characteristics of blades were not addressed.

5 Conclusions

To date, WEICan has shown with some certainty that some small wind turbines are underperforming due to optimistic power curves and AEP predictions. These two issues are likely caused by unreliability, poor siting or poor turbine design. These conclusions closely mirror the results of the Status Report on Small Wind Energy Projects Supported by the Massachusetts Renewable Energy Trust [4] and the Warwick Trials [5].

Economically, given the large initial cost and continuous maintenance costs, small wind as a distributed renewable energy source has become extremely unattractive from a financial point of view. If the plummeting cost of photovoltaics is considered along with their near zero maintenance requirement, small wind cannot compete in most markets.

Acknowledgements Part of this work was funded by the Saskatchewan Research Council's Small Wind Study.

References

1. Orrell, A., Foster, N.: 2015 distributed wind market report. US Department of Energy, Richland (2016)
2. IEC: IEC Standard 61400-2: Wind turbines–Part 2: Small wind turbines Edn 3.0 2013-12. IEC, 2013
3. Fu, R., Chung, D., Lowder, T., Feldman, D., Ardani, K., Margolis, R.: U.S. solar photovoltaic system cost benchmark: Q1 2016. NREL (2016)
4. Shaw, S., Rosen, A., Beavers, D., Korn, D.: Status report on small wind energy projects supported by the Massachusetts Renewable Energy Trust. Cadmus Group (2008)
5. Encraft: Warwick wind trials final report, 2009

[7] http://www.windatlas.ca/

6. James, P.A.B., Sissons, M.F., Bradford, J., Myers, L.E., Bahaj, A.S., Anwar, A., Green, S.: Implications of the UK field trial of building mounted horizontal axis micro-wind turbines. Energy Policy **38**(10), 6130–6144 (2010)
7. Sissons, M.F., James, P.A.B., Bradford, J., Myers, L.E., Bahaj, A.S., Anwar, A., Green, S.: Pole-mounted horizontal axis micro-wind turbines: UK field trial findings and market size assessment. Energy Policy **39**(6), 3822–3861 (2011)
8. Pagnini, L.C., Burlando, M., Repetto, M.P.: Experimental power curve of small-size wind turbines in turbulent urban environment. Appl. Energy **154**, 112–121 (2015)
9. IEC: IEC Standard 61400-12-1: Wind energy generation systems—Part 12-1: Power performance measurements of electricity producing wind turbines. IEC (2017)
10. Wood, D.: Small Wind Turbines: Analysis, Design and Application. Springer, London (2011)

Energy-Yield-Based Optimization of a H-Darrieus Wind Turbine in Skewed Flow

A. Bianchini, F. Balduzzi, G. Ferrara and L. Ferrari

Abstract Recent experiments and theoretical models showed that the aerodynamic performance of H-Darrieus wind turbines can even be enhanced in case of moderate skew angles, which are typical of installations in the urban environment. In this study, a design procedure oriented to the maximization of the annual energy yield in skewed flow, instead of the maximum rated power, was carried out. 14400 test cases of H-Darrieus rotors were simulated with a numerical code based on a Blade Element Momentum approach, including an in-house model to account for the skewed flow, and compared on the basis of their energy-yield capabilities for different annual wind distributions. The analysis highlighted that the optimal design configurations in skewed flow significantly differ from the corresponding ones in case of aligned flow and also that a design oriented to the maximum energy-yield in skewed flow can make H-Darrieus rotors competitive for urban installations in comparison to HAWTs.

Keywords Darrieus · Wind turbine · Skewed flow · Energy yield · BEM

List of Symbols and Abbreviations

A	Area (m^2)
a	Induction Factor
AR	Aspect Ratio
BEM	Blade Element Momentum

A. Bianchini (✉) · F. Balduzzi · G. Ferrara
Department of Industrial Engineering (DIEF), Università degli Studi di Firenze, Via di Santa Marta 3, 50139 Florence, Italy
e-mail: bianchini@vega.de.unifi.it

L. Ferrari
Department of Energy, Systems, Territory and Construction Engineering (DESTEC), University of Pisa, Largo Lucio Lazzarino, 56122 Pisa, Italy
e-mail: lorenzo.ferrari@unipi.it

L. Battisti and M. Ricci (eds.), *Wind Energy Exploitation in Urban Environment*, Green Energy and Technology, https://doi.org/10.1007/978-3-319-74944-0_12

c	Blade Chord (m)
c_P	Power Coefficient
D	Turbine Diameter (m)
F	Force (N)
f	Frequency
H	Turbine Height (m)
HAWT	Horizontal Axis Wind Turbine
m	Mass (kg)
N	Blades/Struts Number
P	Power (W)
R	Turbine Outer Radius (m)
Re	Reynolds Number
T	Annual Time of each Wind Class (h)
T_{par}	Parasitic Torque of the Struts (Nm)
t	Airfoil Thickness (m)
TSR	Tip-Speed Ratio
u	Wind Class (m/s)
U	Wind Speed (m/s)
U_∞	Wind Speed at Infinite (m/s)
$\bar{U}$	Average Wind Speed (m/s)
VAWT	Vertical Axis Wind Turbine
w	Relative Wind Speed (m/s)

Superscripts

$*$	Per Unit Area
$\rightarrow$	Vectorial Quantity

Subscripts

air	Air
b	Blade
en	Energy
lim	Limit
res	Resistant Component
ST	Struts

Greek Letters

Φ	Turbine Shape Factor
γ	Skew Angle (deg)
η_{en}	Energy-conversion Efficiency
ϑ	Azimuthal Angle (deg)
ν	Kinematic Viscosity (m^2/s)
ξ	Chord/Diameter Ratio
ρ	Density (kg/m^3)
σ	Solidity
ς	Structural Stress (N/m^2)
ω	Rotational Speed (rad/s)

1 Introduction

Even though the majority of the wind energy capacity presently installed worldwide is provided by large wind farms comprised of horizontal axis wind turbines [1], increasing interest is received by smaller rotors in view of a delocalized power production [2]. Typical examples of such installations are represented by turbine integration in smart energy grids [3] or wind energy use to satisfy the demand of off-grid communities [4] or remote users [5]. Installed capacity comprised of smaller turbines, both on- and off-grid, is growing particularly in the U.S. [6] with similar trends notable in other developed countries. Research into novel small designs is also on the increase.

Building integration forms a large part of this, with studies looking at the challenges presented by the pre-existing built environment [7–9] or aiming at designing new buildings that produce favorable conditions for wind energy production [10].

Darrieus-type vertical axis wind turbines (VAWTs) have been identified as suitable for small-scale installations due to their mechanical simplicity and ease of installation and maintenance afforded by the positioning of generation equipment at ground level [11]. They are also particularly suited to urban installations due to their good handling of turbulent and unstructured flows, with low noise emissions and high reliability [12, 13]. In particular, recent experimental studies [14, 15] and theoretical models [16, 17] have demonstrated that—differently from HAWTs [18]—Darrieus turbines can even enhance their efficiency in case of moderately skewed flows (up to a 20% increase of the power coefficient, depending on the turbine design), which are typical of urban installations [10, 19] and of floating platform functioning [20], which indeed represents the other promising application for these rotors.

The physical reasons underlying this behavior are multiple. As first discussed by Ferreira et al. [16], even though—analogous to HAWTs—the effective wind speed on the blade is reduced by the cosine of the skew angle, for moderate skew angles (up to 30°–40°) the area swept by the turbine is increased in reason of the arising contribution from the discovered downwind part of the virtual cylinder swept by the blades. As will be illustrated in Sect. 2, it was also shown by Bianchini et al. [17] that the flow conditions in different parts of the machine are altered by an oncoming skewed flow, generating a larger energy extraction in the downwind region.

Based on the above, it can be argued that dedicated design studies aimed at optimizing a Darrieus rotor for an attended non-null angle of skew may induce an improvement of the energy-conversion efficiency of these machines, making them more competitive with HAWTs, which conversely suffer from a strong efficiency reduction in similar conditions. In the present study, the results of a multivariate analysis are reported, whose main aim was to optimize the proportions of a small H-Darrieus VAWT for use in installations having different mean skew angles and attended wind distributions.

2 Methods

2.1 Energy-Yield Based Design Approach

Due to marketing reasons, the majority of commercial rotors are generally designed and optimized for a specific wind speed (e.g. the speed which ensures the highest energy production), but the rated power values, which are often perceived by the final customer as the most valuable indicator of the "quality" of the product, are declared for their nominal wind velocity, i.e. the highest functioning speed, which provides the maximum power production. Although the importance of accounting for energy instead of power does not come as a surprise to researchers, the implications of this theoretical dichotomy are quite often not completely understood by the final customer of small and medium wind turbines [21]. In further detail, the actual operating conditions of a rotor over a year can indeed be very far from the nominal one [22]; the available wind energy indeed can be concentrated at the lower wind velocities of the yearly distribution in the installation site, which can be often correctly approximated by a Weibull function [23]; in addition, the specific features of the final environment (e.g. local accelerations, effects of obstacles, etc.) are very important in determining the real characteristics of the flow which invests the rotor (e.g. [24]). As a result, a turbine optimized only for one (generally quite high) wind speed could provide poor performance during the largest part of its operating time, with a remarkable reduction of the energy produced and, consequently, of the convenience of the investment [19]; this effect is maximized whenever the shape parameter of the distribution is low, i.e. when the higher wind velocities give a remarkable contribution to the global energy harvesting, although the mean velocity is quite low.

A design approach for small Darrieus turbines based on the maximization of the annual energy-yield (i.e. the sum of the energy contributions at all the wind speeds experienced over the year) was then recently proposed by Bianchini et al. [25]. The same approach is also proposed in the present study and detailed below. More specifically, the goal of the design process was to define the overall dimensions of a turbine having a given swept area of 4 m^2 (typical of small VAWTs) able to ensure the maximum energy harvesting in a yearly horizon as a function of the attended wind distribution in the installation site and the skew angle (assumed constant for all the wind speeds). With this goal in mind, 14400 test cases, i.e. permutations of a specific geometrical configuration (100 cases), a wind distribution (6 cases), load system (3 cases) and a skew angle (8 cases), were tested and analyzed by means of a specifically developed numerical code based on an improved Double Multiple Streamtubes Blade Element Momentum (BEM) approach and able to account for the skew angle effects.

2.2 *Wind Distributions*

According to the original study of Bianchini et al. [25] for a conventional aligned flow, six annual wind profiles were considered during the design process.

More in detail, the general shape of a Rayleigh distribution (i.e. a Weibull distribution having a constant shape factor equal to 2.0 [26]) was selected, whereas the scale factor of the same was modified in each case in order to match an average wind speed ($\bar{U}$) increasing by one from 3 to 8 m/s. A constant cut-out velocity of 16 m/s was imposed for safety reasons to all the tested rotors [25], based on the authors' experience in small industrial Darrieus rotors. On the other hand, the cut-in speed was directly calculated with the aerodynamic code for each rotor on the basis of its aerodynamic behavior at low wind conditions: a variable cut-in speed between 2 and 3 m/s was measured in the tested rotors.

The wind profiles investigated in this study are shown in Fig. 1, in which the cut-in and the cut-out limits are also displayed.

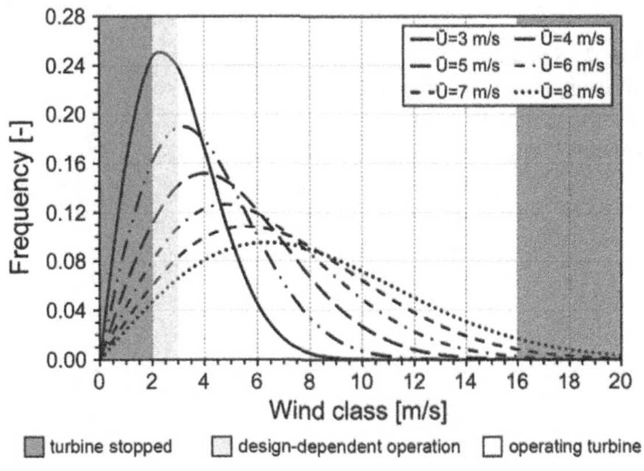

Fig. 1 Considered wind distributions and turbine operating ranges

2.3 Design Variables and Test Plan

Since the possible design choices for a Darrieus rotor can be indeed very many, a preliminary selection of some of them was needed in order to let the attention be focused on the effects of a skewed flow. The following main assumptions were made:

- the H-Darrieus configuration (i.e. straight blades) was selected for all the rotors (see Fig. 2) with a constant blade number $N = 3$, since it actually represents the best compromise in terms of efficiency and structural solidity [27];
- a constant swept area of $A = 4\ \mathrm{m}^2$ was considered. Such a value is somehow representative of small urban wind turbines in terms of overall dimensions and rated power (approximately 2 kW @ 14 m/s);
- blades were made of a NACA0018 airfoil along the entire span;
- two supporting struts for each blade were applied, whose dimension was varied as a function of the calculated structural stresses due to both the centrifugal component and the aerodynamic loads [25].

Regarding the airfoil, it must be noticed that in the present study the selection of the NACA0018 profile must be intended by the reader net of the virtual camber correction [28]. In further detail, it is here assumed that each airfoil actually perform like the NACA0018 in straight flow; as a result, the different turbines should be manufactured in each case using conformally transformed airfoils (based on the chord-to-radius ratio of the rotor) with inward concavity [29].

Some significant design parameters to be analyzed as a function of the attended skew angle were then chosen. According to the proposal of [21] and [25], the attention was focused on:

- the height/diameter ratio ($\Phi = H/D$);
- the chord/diameter ratio ($\xi = c/D$);
- the struts dimensions and shape.

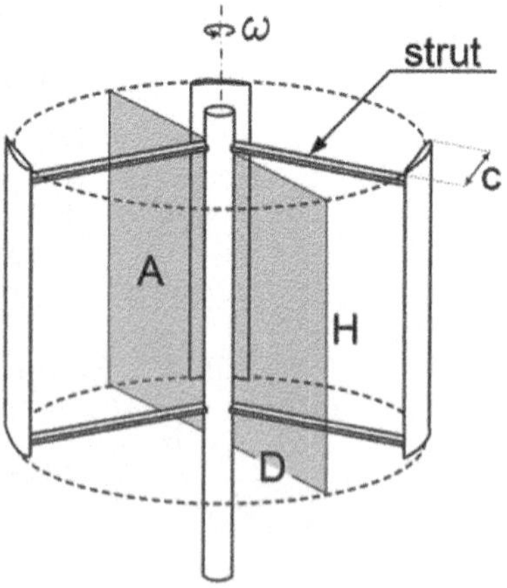

Fig. 2 Schematic description of a H-Darrieus rotor including the main design parameters

The aerodynamic implications of these parameters have been discussed extensively by [25] and are only briefly summarized here.

The Φ ratio has a double effect in defining the features of the rotor. First, it can be considered as a "shape factor" of the turbine, i.e. an indicator of the visual proportions of the cylinder swept during the revolution. On the other hand, it is quite relevant in defining the "aerodynamic requests" of the rotor, i.e. for a fixed swept area, low Φ values are typical of a machine which needs to maximize D in order to increase the peripheral speed. Conversely, high Φ values can be related to machines in which the efficiency of the blade (i.e. high blade Aspect Ratios—Eq. 1) is preferred.

$$AR_b = \frac{H}{c}. \tag{1}$$

The ξ ratio is instead an indicator of the solidity (σ) of the rotor, being (Eq. 2):

$$\sigma = \frac{Nc}{D}. \tag{2}$$

Moreover, the ξ ratio can also suggests how the Reynolds number (Re_b—Eq. 3) needed for an efficient behavior of the airfoils is about to be obtained on the blades: high values of ξ indeed indicate that the chord length is increased, whereas low ξ values can be related to rotors in which the relative wind speed is increased by means of an increase of the relative wind speed on the airfoil [21], as readily arguable upon examination of conventional Eq. 4 applied in BEM models.

$$\mathrm{Re}_b = \frac{wc}{\nu}. \tag{3}$$

$$w = \vec{\omega}R + \vec{U}\,(1 - a)\,. \tag{4}$$

The struts' dimension and shape is finally of major importance for small H-Darrieus rotors, since their parasitic torque may substantially reduce the power output [27]. Moreover, in case of skewed flow, they are subject to a non-null incidence, then they even more directly contribute to the overall performance of the turbine.

Stated the above, Table 1 reports the test plan of the present study.

Finally, focusing on the boundary conditions in terms of loads applied to the rotor, the three different configurations proposed by Bianchini et al. [25] were again analyzed:

- **CASE 1—Aerodynamics only**: this configuration actually refers to an hypothetical solution of a fully aerodynamic relationship between the geometrical features of the rotor and the power performance (i.e. neither structural limitations nor the parasitic torque of the struts are included). Although not feasible in reality, Case 1 results are particularly interesting since they allow one to define the aerodynamic dependence on the different design parameters.
- **CASE 2—Centrifugal load**: a limitation on the centrifugal stress acting on the blades was added as a function of the optimal design rotational speed.

Table 1 Test plan of the present study

Parameter	Value/range
$\Phi = H/D$	0.2–0.4–0.6–0.8–1.0–1.2–1.4–1.6–1.8–2.0
$\xi = c/D$	0.005–0.010–0.015–0.020–0.040–0.060–0.080–0.100–0.150–0.200
A	4.0 (m^2)
Airfoil	NACA0018
$\bar{U}$	3–4–5–6–7–8 (m/s)
Constraints	aerodynamics only—centrifugal load—struts' parasitic torque
Skew angle (γ)	0–5–10–15–20–25–30–35 (deg)

- **CASE 3—Strut parasitic torque**: in this latter configuration, both a limitation on the centrifugal stresses and strut dimensions variable as a function of the rotational speed were included. The resistant torque of the struts was also taken into account.

The structural characteristics of the rotors can indeed be extremely variable in dependence on the design, the material and the manufacturing technology. A multivariate analysis on these aspects is outside of the scopes of the present study. In order to provide a realistic estimation of the structural loads, however, a manufacturing technology based on extruded aluminium blades with a hollowed section was considered, utilizing real data on both the maximum centrifugal stress and the maximum mechanical stress on the struts available from previous design experience [30]. In detail, in Case 2 the cut-out speed of the turbine was calculated in each case-study based on the hypothesis that the maximum centrifugal stress at the middle of the blade (where the maximum displacement is located) would not exceed a fixed stress limit (Eq. 5).

$$\varsigma_b = \frac{F_b}{A_{res\_b}} = \frac{m_b \omega^2 R}{A_{res\_b}} \leq \varsigma_{\lim\_b} \tag{5}$$

where A_{res_b} is a conventional resistance area which takes into account the structure of the stiffeners inside the airfoil. In particular, all the airfoils were reproduced with a hollowed section having a constant skin thickness (3 mm) and three, equally spaced, rectangular stiffeners perpendicular to the chord of variable thickness; moreover, the trailing edge was considered as solid with a fillet radius increasing proportionally to the chord of the blade (Fig. 3).

Fig. 3 Hypothesized blade manufacturing solution

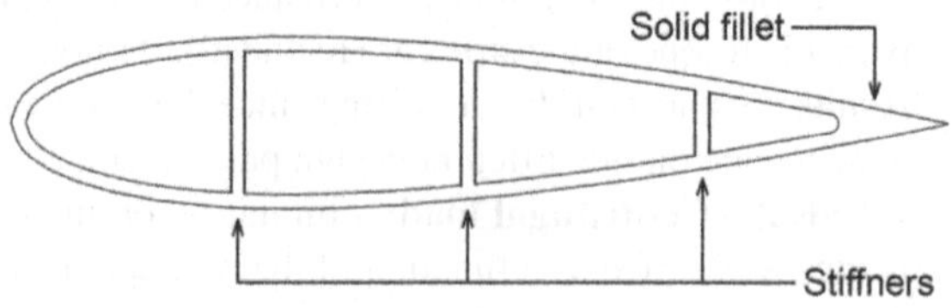

In addition, in Case 3 even the dimensions of the two struts per blade were modified from one configuration to another in order to satisfy the stress limit of the blades due to the centrifugal force (Eq. 6).

$$\varsigma_{ST} = \frac{F_{ST}}{A_{res\_ST}} = \frac{1}{N_{ST}} \frac{m_b \omega^2 R}{A_{res\_ST}} \leq \varsigma_{\lim\_ST} \tag{6}$$

It is worth remarking that the load case considered in Eq. 6 takes into account only the tensile stresses generated by the centripetal acceleration of the rotor. In the authors' experience [31], this load condition is quite realistic for small-size rotors (i.e. with a swept area up to 4–5 m^2). In this work, the bending stresses were neglected but a more accurate description of their influence will be carried out in future works.

Finally, regarding the strut shape, a constant design was figured out, corresponding for simplicity to the same airfoil (i.e. NACA0018) of the blades.

2.4 Simulation Code and Skewed Flow Model

Performance simulations were carried out with the VARDAR code of the Department of Industrial Engineering of the University of Florence. The code has been specifically developed for H-Darrieus wind turbines using an improved version of a *Double Multiple Streamtubes Approach with Variable Interference Factors* [11, 32] embedded with some advanced routines to account for the main secondary and parasitic effects [30], including different dynamic stall models [33] and tip-losses. The prediction capabilities of the VARDAR code have been validated with several test campaigns in the wind tunnel on 1:1 models of H-Darrieus rotors. The code was able to correctly predict both the power curves at different wind speeds and the starting ramps of rotors with two or three blades, either straight or helix-shaped. Further details on the code, its validation and main features can be found by the reader in [21, 28, 30, 33]. In particular, the code includes two original sub-models which were pivotal for the present analysis. The first one is the model for the performance analysis in skewed flow conditions, that has been validated by some of the authors against experimental data in [15, 17], obtaining notable agreement as testified by Fig. 4 which reports the comparison of model predictions with the data of Ferreira et al. [16] (Fig. 4a) and Bianchini et al. [17] (Fig. 4b), respectively. For brevity reasons, all the theoretical assumptions of the model and their implementation within the BEM theory cannot be reported here but can be found by the reader in [17].

It is however mandatory to briefly summarize here the mail model characteristics:

- the aerodynamic performance of the rotor in skewed flow is directly simulated (no a posteriori correction);
- the cross-flow principle is applied [34], i.e. the wind speed component which actually contributes to the creation of the lift and drag forces on the airfoils is the

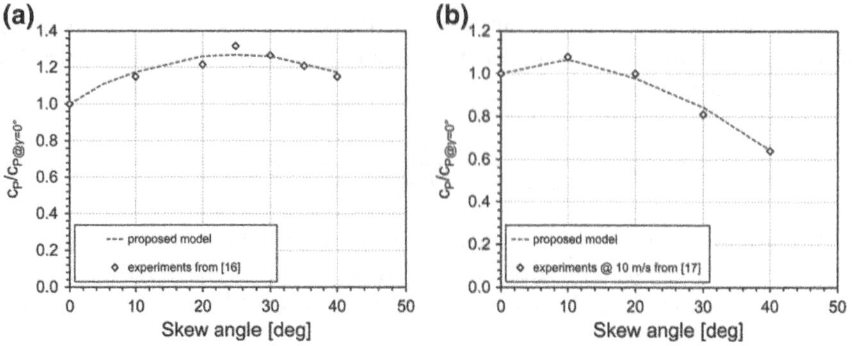

Fig. 4 Validation of the theoretical model for skewed flow conditions against: **a** experiments by [16]; **b** experiments by Bianchini et al. [17]

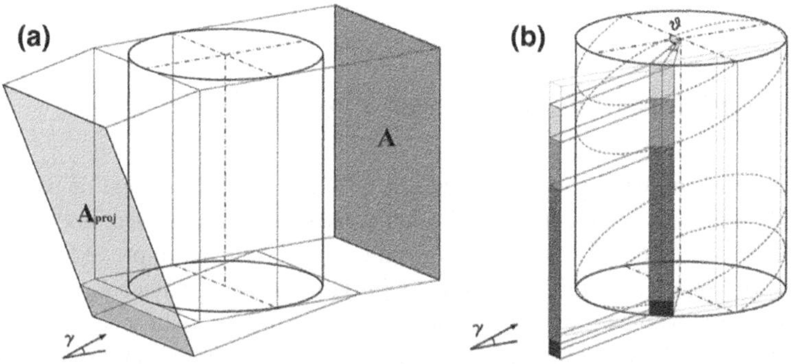

Fig. 5 Theoretical turbine schematization in the skewed flow model (from [17])

only projection of the absolute wind speed on the direction orthogonal to the blade span (i.e. a function of the cosine of the skew angle);

- the increase of the turbine equivalent swept area is considered as in Fig. 5a;
- the turbine is divided into five sectors, in each of which the blade-flow interaction changes as a function of the azimuthal position, i.e. the turbine works either in a multitubes-like or a double-multitubes-like way (Fig. 5b).

The second important module of the VARDAR code was represented by the routine for the calculation of the resistant torque of the struts. Indeed, some of the authors recently developed a model for the estimation of this contribution in case of aligned flow [35], which provided a very accurate estimation in comparison to experiments. In case of a skewed flow, however, the model needed a further update since the skew angle may in fact lead to a non-negligible contribution (even a positive one) by the airfoil-shaped struts. To this end, the original model—in which the induced aerodynamic drag of the strut section during the revolution is thought to be given by the projection of the relative speed (variable with the radius and calculated with a multitubes BEM approach for each azimuthal position) on the tangential

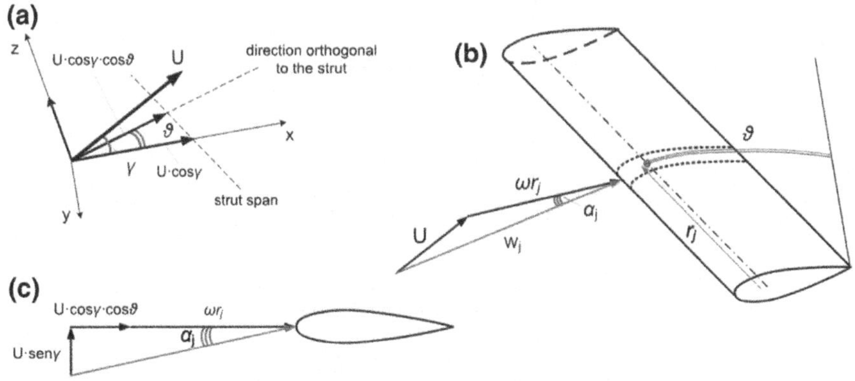

Fig. 6 Theoretical model for the calculation of the torque of the struts in skewed flow

direction [35]—was here modified to reconstruct in each azimuthal position the velocity triangle on the strut at different span sections of the same (Fig. 6a). Under these hypotheses, the incidence angle on the strut (Fig. 6b)—at which the lift and drag coefficients of the strut itself are considered—is given for a generic azimuthal position ϑ and span section j by Eq. 7 (Fig. 6c):

$$\tan\alpha_j = \frac{U\ \sin\gamma}{\omega r_j + U\ \cos\gamma\ \cos\vartheta} \tag{7}$$

This new model was also validated against experimental data. In particular, the results of [15] were used, where the struts of a H-Darrieus turbine, made exactly of a NACA0018 airfoil, were tested in the wind tunnel for different skew angles. As an example, Fig. 7 reports the comparison between the measured torque (reported in a dimensionless form to preserve the non-disclosure agreement with the industrial partner of the activity) of the six struts (two per blade)—in case of either aligned or skewed flow ($\gamma = 10°$)—and the numerical model for the same skew angle. Upon examination of the figure, good agreement can be appreciated between simulations and experiments. In particular, the new model was able to properly estimate the mitigation of the parasitic torque in case of a moderate skew angle in comparison to the original values in aligned flow. Finally, it has also to be noted that, in case of a skewed flow, lower struts (struts are considered applied to the blade ends) partially shadow a portion of the blade span. For simplicity, it was here considered that the active span of the blades is directly reduced by the projection of the strut length on the span at each azimuthal position.

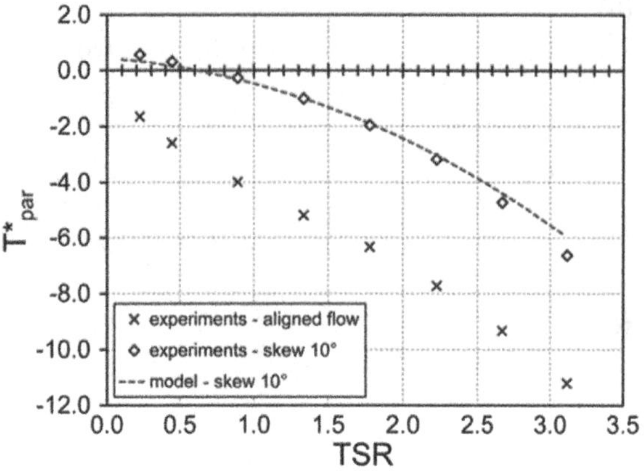

Fig. 7 Validation of the new model for the resistant torque of the struts in skewed flow

2.5 Data Reduction

The VARDAR code provided the attended power coefficient of each configuration at all the wind speeds within the rage from the cut-in (variable) to the cut-out (fixed) limits.

The rated power coefficient at each wind speed was conservatively evaluated in correspondence with the calculated performance 0.2 points of TSR after the peak of the operating curve [25]. It is also worth noticing that the cut-in speed was set to the wind speed for which a positive power coefficient was obtained. For each wind distribution in the site (i.e. for each average wind speed), performance maps were created [21], each of which contained the overall energy conversion efficiency (η_{en}) of the specific rotor, defined as in Eqs. 8 and 9.

$$\eta_{en} = \frac{\sum_{u=cut-in}^{cut-out} c_P(u) \cdot u^3 \cdot T(u)}{\sum_{u=cut-in}^{cut-out} u^3 \cdot T(u)} \tag{8}$$

$$T(u) = f(u) \cdot 8760. \tag{9}$$

As one may notice, the energy conversion efficiency (η_{en}) was defined as the ratio between the annual energy yield of the turbine over the year (i.e. the real extracted power at each wind speed multiplied by the time fraction—in hours—during which that wind blows *T(u)*) and the theoretical energy contained in the wind itself. Based on its definition, η_{en} differs from the classical power coefficient and allows one to simultaneously take into account both the efficiency of the turbine at all the wind speeds expected over the year and the effects related to the starting and resistance capabilities of the rotor (due to the variable cut-in and cut-out velocities considered).

A typical energy efficiency map is shown in Fig. 8 (e.g. $\bar{U} = 6$ m/s, $\gamma = 5°$, Case 3).

A linear interpolation was performed between the calculated points (see Table 1) in order to more sharply outline the contours. An optimum-design zone (white-colored in the figure) can be typically distinguished in the bottom side, where the

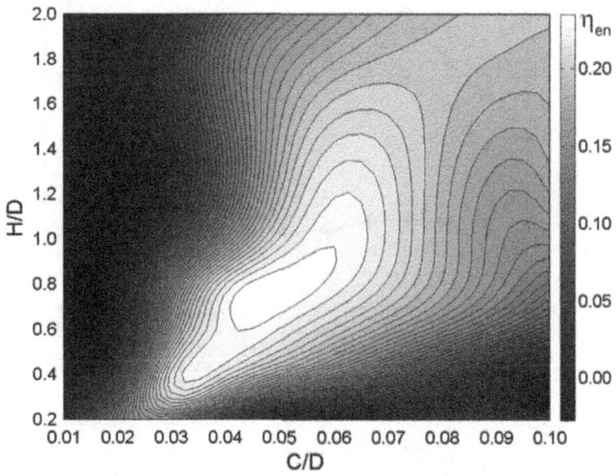

Fig. 8 Sample energy efficiency map ($\bar{U} = 6$ m/s, $\gamma = 5°$, Case 3)

combination of the Φ and ξ parameters ensures the best compromise in terms of functioning Reynolds numbers on the airfoil (high peripheral speed and chord) and efficiency of the blades (high Aspect Ratio). Moreover, one can also notice that the efficiency is almost zero in the left side of the map, where the very small chords remarkably reduce the lift generation, and becomes lower also in the right-bottom corner of the map, where the turbine height tends to zero.

Within each map, a numerical identification of the maximum was performed, with the constraints of neglecting design solutions which imply blade Aspect Ratios lower than 5 (too high tip effects, not adequately modeled by the code) or higher than 35 (too slender blades [25]). This procedure finally led to the definition of the geometrical features of the rotor able to provide the best energy harvesting over a year for the attended load case, average wind speed and skew angle in the site.

3 Results and Discussion

The variation trends describing the dependence of the optimal values of the most relevant design parameters on the average wind speed for the investigated configurations in Case 1 are presented in Fig. 9. In the figure, the left-hand side charts refer to an aligned flow condition ($\gamma = 0°$), while the right-hand side ones report the results in case the oncoming wind is inclined by a constant skew angle $\gamma = 10°$ with respect to a plane orthogonal to the rotor axis. For brevity reasons, this latter condition was here taken as representative of the modifications induced by the skewed-flow conditions on the turbine design for each load case.

Focusing first on the baseline case with no skew, it can be noticed that the same trends originally described by [25] were here confirmed. In particular, when the average wind speed in the site increases, the best solidity decreases constantly and the Aspect Ratio rises significantly. Focusing on the dimensional parameters, the

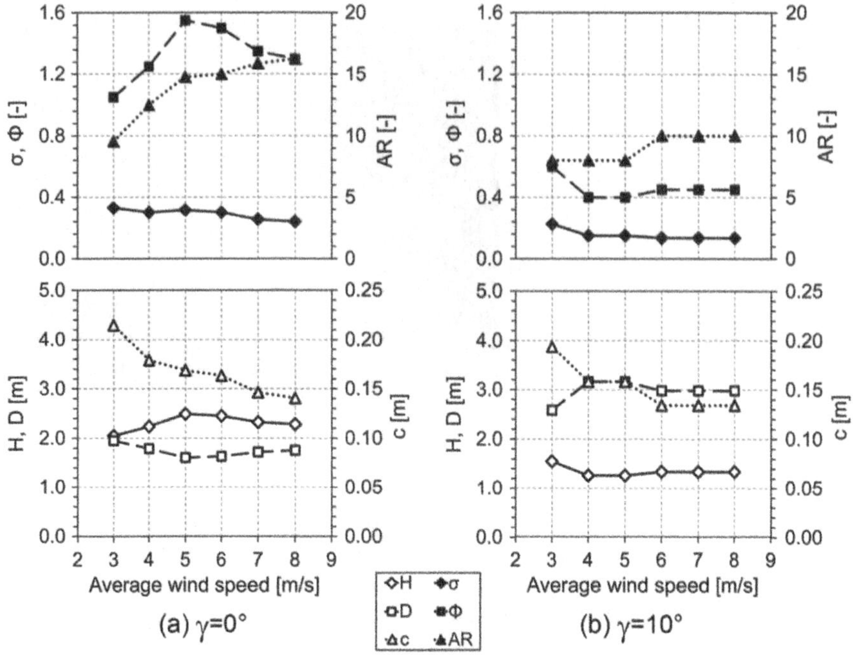

Fig. 9 Optimal trends of the relevant design parameters of the turbine as a function of the average wind speed in the site in Case 1: a comparison between designs in aligned or skewed ($\gamma = 10°$) flow

blade chord decreases by increasing the average wind speed (with a steeper trend for medium-low $\bar{U}$ and a flatter trend with an increased $\bar{U}$), whereas the turbine diameter and height have an opposite trend with a decrease and an increase in the first part, respectively, followed by a fairly constant trend in the second part. Analyzing the aerodynamic implications of these trends, it is worth pointing out that, when the average wind velocity experienced by the turbine is low (left side of the plots), the best design is that ensuring the highest local Reynolds numbers on the blade by means of the geometrical proportions. As a consequence, the chord values are maximized (Eq. 3) at low wind speeds and then the solidity values are high (Eq. 2). By increasing the average wind speed, the relative velocities experienced by the blades are increased, mitigating the dependence on the Reynolds number, even if the chords are reduced. For the highest average wind velocities, the further increased Reynolds number allows a reduction of the diameter, coupled with a simultaneous increase of the turbine height, which ensures higher Aspect Ratios ($AR = H/c$), i.e. more efficient blades [11].

Focusing now on the same trends in case of a skewed flow, it is first apparent that they are generally less pronounced. The overall decrease of the solidity—mainly connected to the reduction of the chord—as a function of the average wind speed is still distinguishable, confirming the strong dependence of the aerodynamic on the local Reynolds number on the blades. However, it can be noted that as soon as

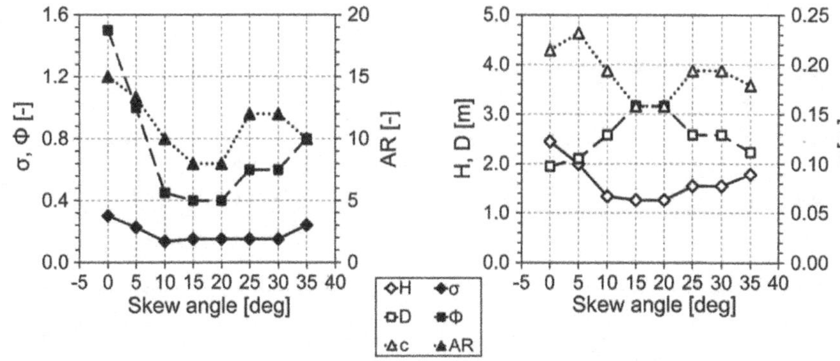

Fig. 10 Optimal trends of the relevant design parameters of the turbine as a function of the skew angle in Case 1 ($\bar{U} = 6$ m/s)

the Reynolds number implications become less demanding (approximately starting from average wind speeds higher than 6 m/s), the optimal design solution suddenly becomes independent on the average wind speed. In these conditions, the optimal design is defined by the maximization of the skewed flow effects. In particular, comparing the optimum design configurations with their homologous ones in aligned flow, it can be noticed that:

- larger diameters are preferable, since they maximize the increase of the swept area (see Fig. 4), i.e. the wind flow "captured" by the turbine [16];
- the upper limit in the maximization of the diameter is of course imposed by a balance with the blade efficiency loss due to tip-effects (function of the blade Aspect Ratio), which sets out the minimum blade height;
- for a given average wind speed, the larger diameters increase the peripheral speed and then the blade Reynolds numbers with respect to the same cases in aligned flow. As a result, chords can be generally shorter, mitigating the problems in terms of reduced blade AR;
- as a consequence of the above, the optimal rotors in skewed flow are generally characterized by lower solidities with respect to the aligned-flow cases.

If the above comparison of design trends between aligned and skewed flow conditions can be considered of general validity, it is also interesting to analyze the dependence of the same parameters on the skew angle, reported in Fig. 10. Upon examination of the figure, it is apparent that:

- the modifications induced by the skewed flow strongly affect the design of the rotor as soon as the skew angle reaches approximately $\gamma = 10°$;
- beyond that value, the optimal design remains more or less constant (even if the discussed trends are still slightly visible). This is again due to the fact that a further diameter increase to exploit the skew angle would dramatically reduce the AR of the blades (the reader has to remember that the optimization was carried out for a fixed swept area);

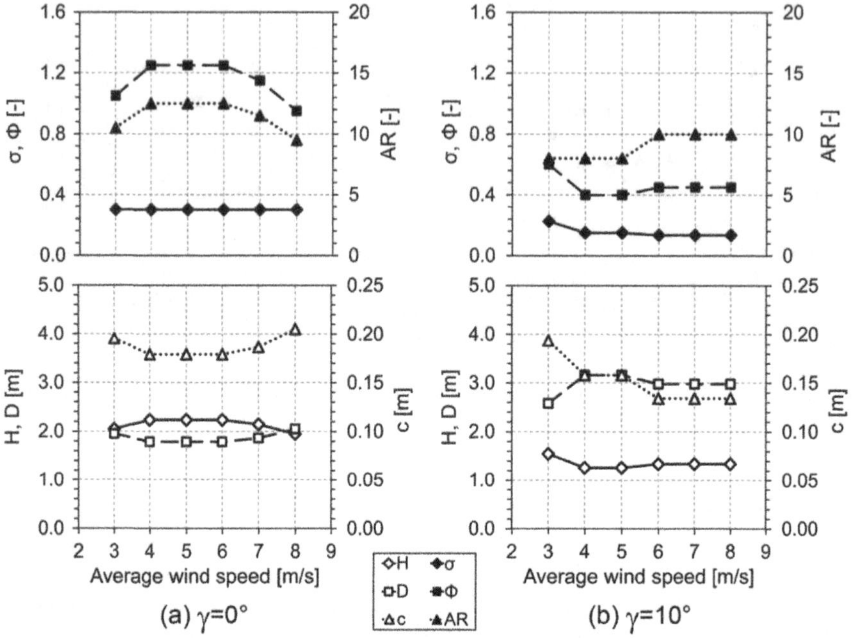

Fig. 11 Optimal trends of the relevant design parameters of the turbine as a function of the average wind speed in the site in Case 2: a comparison between designs in aligned or skewed ($\gamma = 10°$) flow

- on the other hand, when the skew angle becomes too large (i.e. approximately beyond $\gamma = 30°$), the trends are suddenly inverted. In this condition, the reduction of the relative speed on the blades induced by the decreased active component of the wind speed [16] indeed again poses problems connected to the Reynolds number conditions and then to blade efficiency. Moreover, as discussed by [16], if the Φ ratio is too low, part of the wind can pass through the rotor without interacting with the blades;
- as a result, for high skew angles, the optimal Φ ratios re-start increasing (in reason of an increase of the turbine height) together with blade AR, ensuring a recovery of the aerodynamic efficiency of the same. The diameters are accordingly slightly reduced.

The results obtained from Case 1 are very useful to comprehend the aerodynamic trends; structural constraints like the centrifugal load have, however, relevant impact on the best design compromise of a machine. To this end, Fig. 11 reports the results of the investigation on the study-cases in Case 2, in which a ς_{lim_b} (Eq. 6) of 90 N/mm^2 was considered [17]. As already noticed by [25], the introduction of a limit on the centrifugal stresses (i.e. on the combination of the peripheral speed of the rotor and the blade mass, mainly dependent on the chord) remarkably modifies the optimal design of the turbine.

Focusing first on the aligned-flow case, it can be noticed that, when the average wind velocities are low (i.e. 3–4 m/s), the structural constraints slightly affect the definition of the best design parameters with respect to Case 1; the operating rotational speeds at these wind velocities are indeed low and the resulting centrifugal loads do not exceed the structural limits. As soon as the average wind speed increases, however, the reduction of the solidity with the average wind speed noticed in Case 1 ceases (an almost constant value is reached). To respect the limits on the centrifugal stresses, the revolution speed of the turbine has to be decreased (the centrifugal stresses have indeed a quadratic dependence on the revolution speed). To do so, the optimal chords increase once again together with the diameter: the best compromise in terms of energy-yield capabilities in fact comes from a reduction of the peak efficiency of the turbine (high solidity and lower AR) which, however, ensures a reduction of the operating rotational regime [11, 27] and then a mitigation of centrifugal loads.

Focusing instead on the trends in case of a skewed flow, it is apparent that in this case the introduction of a limit on the centrifugal stresses did not modified the design configurations highlighted in Case 1. As discussed, the effects of the skewed flow had already modified the shape of the rotor in the sense of an increase of the diameter to exploit the increased swept area of the machine: the optimal configurations were then already characterized by lower optimal tip-speed ratios.

Moreover, the centrifugal stresses in case of a skewed flow can also take advantage from a general reduction of the chords (i.e. of the blade mass) allowed by both the larger diameters and the need of partially recover the blade efficiency through an increase of the AR.

According to Case 1, Fig. 12 finally reports the variation of the optimal design ad a function of the skew angle. Even if the absolute values are slightly modified with respect to Case 1 (e.g. for $\bar{U} = 5$ m/s), the same considerations are still valid here, highlighting like—in a general scenario where the diameters are increased to ensure the maximum exploitation of the skewed flow effects in terms of major wind captation—for higher skew angles, the optimal Φ ratios re-start increasing (in reason of an increase of the turbine height) together with blade AR, ensuring a recovery of the aerodynamic efficiency of the same.

Focusing finally on a real-type machine, a further constraint must be included in the identification of the best design trends: the parasitic torque of the struts is, in fact, a key element in defining both the peak efficiency of the machine and its functioning behavior at different wind speeds. In Fig. 13 the optimum design trends in Case 3 are then presented.

Upon examination of Fig. 13, some interesting remarks on the effect of the parasitic torque can be made. Focusing first on the aligned-flow condition, remembering that the parasitic torque generated by the rotating structures in this case does not contribute to the torque and has a quadratic dependence on the rotational speed of the rotor and a lighter and more complex dependence on the wind velocity, one can notice that:

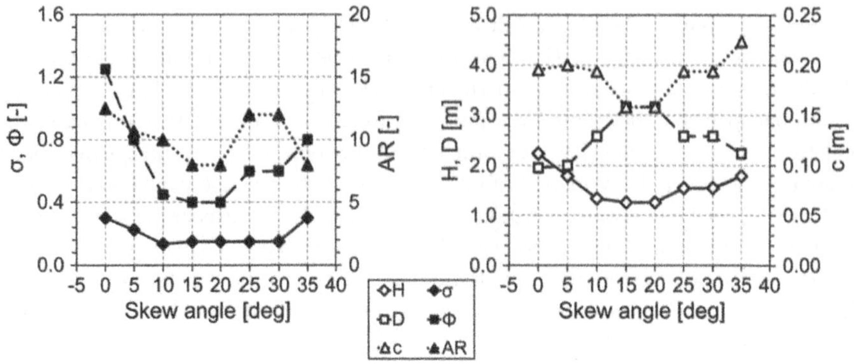

Fig. 12 Optimal trends of the relevant design parameters of the turbine as a function of the skew angle in Case 2 ($\bar{U} = 6$ m/s)

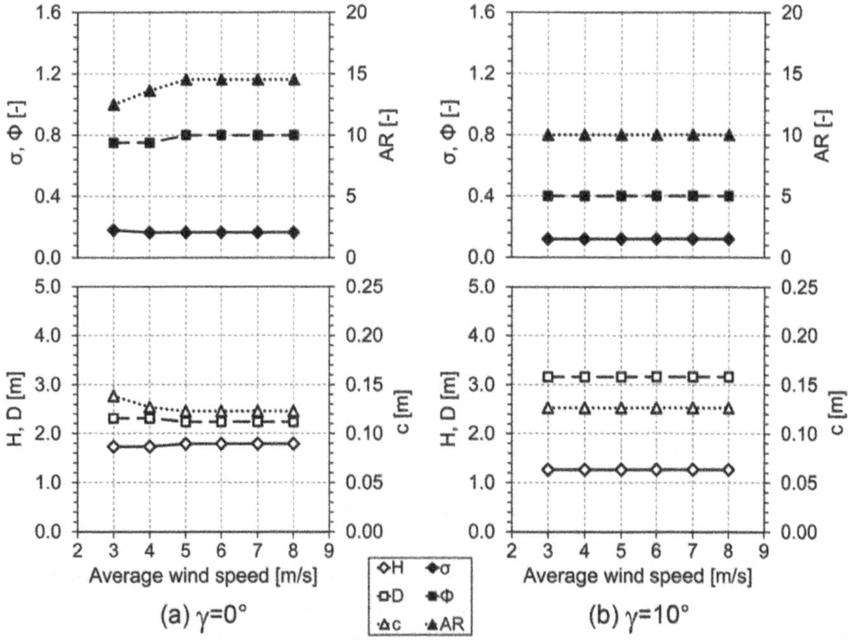

Fig. 13 Optimal trends of the relevant design parameters of the turbine as a function of the average wind speed in the site in Case 3: a comparison between designs in aligned or skewed ($\gamma = 10°$) flow

- the trends of the design parameters as a function of the average wind speed are now flattened. The best design solutions collapse in those configurations which minimize the contribution of the parasitic components;
- as a result, optimal geometries are characterized by higher diameters with respect to Case 2 (and, consequently, lower heights) in order to reduce the operating

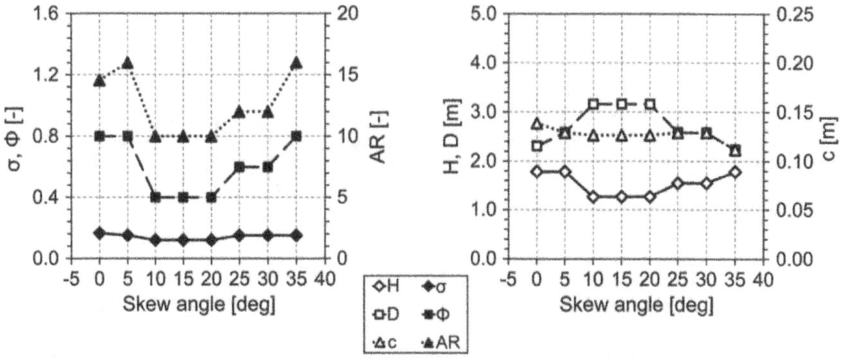

Fig. 14 Optimal trends of the relevant design parameters of the turbine as a function of the skew angle in Case 3 ($\bar{U} = 6$ m/s)

revolution speed of the rotor by means of an increase of the peripheral speed (see Eq. 4).

Moving now on the skewed-flow case ($\gamma = 10°$), it is readily noticeable that the application of all the constraints lead to the definition of a rotor geometry which is definitely independent on the average wind speed. As discussed, diameters larger than those expected in aligned flow are preferable, since they maximize the increase of the wind flow "captured" by the turbine; this tendency—yet discussed for both previous load cases—is even more marked here, since the airfoil-shaped struts are even able to provide a slightly positive contribution to the final torque in some cases. Once again, it is here hypothesized that the upper limit in the maximization of the diameter is imposed by a balance with the blade efficiency loss due to tip effects (function of the blade Aspect Ratio), which defines the minimum values for the blade height and chord length.

Focusing on the variation trends as a function of the skew angle (Fig. 14), the overall trends are again confirmed, even if a general flattening of the same is again noticeable. It is also interesting to point out that a general reduction of blade chord can be noticed here with respect to Case 2. This trend has to be related to the fact that the strut dimensions are directly proportional to the mass of the blades, i.e. to the chord. By reducing it, the use of thinner struts is therefore enabled, with a direct benefit in terms of parasitic component reduction.

As a final remark on the best design trends, Fig. 15 reports the variation as a function of the attended skew angle of the diameter (which was identified as the leading variable in case of skewed flow) in the three load cases and for two different mean wind speeds. Upon examination of the figure, it is apparent that—for low wind speeds (Fig. 15a)—the use of larger diameters for low skew angles is preferable in order to bypass too low blade Reynolds numbers. Moreover, the introduction of the strut torque makes the optimal diameter increase even further for moderate skew angles, where the positive effect of the airfoil-shaped struts can be maximized.

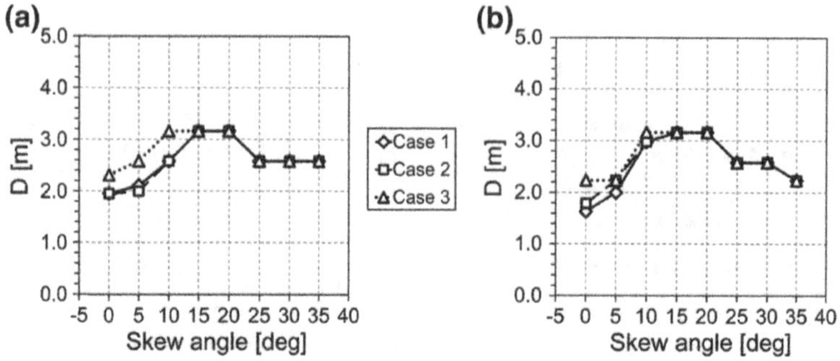

Fig. 15 Optimal turbine diameter as a function of the skew angle in the three investigated load cases: **a** $\bar{U} = 3$ m/s; **b** $\bar{U} = 6$ m/s

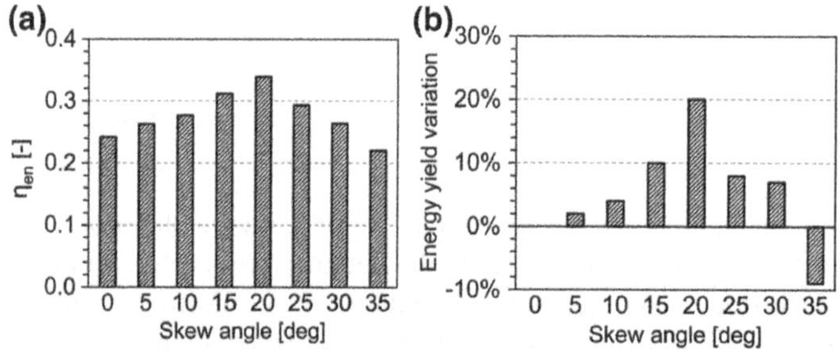

Fig. 16 Average wind speed $\bar{U} = 6$ m/s: **a** efficiency of energy conversion of the most efficient rotor at each attended skew angle; **b** variation of the energy yield between the optimized rotor at each skew angle and the original rotor designed for working in aligned flow

If the results presented so far mainly focused the attention on the design features of the optimal rotors as a function of the attended working conditions (in terms of wind speed and skew angle), focus must be given finally to the real benefits offered by the proposed design strategy in terms of energy yield. It should be remembered that the enhancement of the actual energy production is indeed the only way to make Darrieus rotors somehow competitive with HAWTs. With reference to Case 3 (i.e. the most realistic one, considered fully representative of a real-life design), two different analyses were carried out, and whose main outcomes are reported in Fig. 16.

In particular, Fig. 16a compares—for a representative wind distribution having $\bar{U} = 6$ m/s—the energy conversion efficiency (i.e. the actual energy yield) for the most efficient rotor at each skew angle. It is apparent that the intrinsic capability of H-Darrieus rotors to take advantage from moderately skewed flows (up to 30°) can lead to notable increases of their producibility: for an attended skew angle of 20°, the increase of energy conversion with respect to the aligned flow case was estimated in approximately +40% ($\eta_{en@20^\circ} = 0.339$ vs. $\eta_{en@0^\circ} = 0.242$). Beyond the limit value of $\gamma = 30°$, a decrease of energy conversion was instead noticed (approximately −9%).

Figure 16b conversely reports the difference in terms of actual energy yield between the rotor optimized for an aligned flow and the optimal one at each skew angle. This second analysis is indeed representative on how much a design strategy oriented at maximizing the exploitation of the skewed effects can improve the performance of a H-Darrieus rotor.

It can be noted indeed that, even if also the baseline configuration at $\gamma = 0°$ is able to somehow perform better in skewed flow, a rotor specifically optimized to work at that specific skew angle can provide additional benefits in terms of energy production (up to +20% at $\gamma = 20°$).

Described results are thought of particular interest for future applications of H-Darrieus turbines in the urban environment, with particular reference to rooftop installations, in which the attended skew angle can be reasonably considered constant and a priori estimated upon analysis of the surrounding environment [9].

4 Conclusions

In the present study, a numerical analysis has been carried out to define the potential benefits of a design process for H-Darrieus rotors oriented to the optimization of the annual energy yield as a function of the average wind speed and the attended skew angle in the installation site. In particular, on the basis of six Rayleigh wind distributions with different average wind speeds and seven skew angles (from 0° to 35°), 14400 test cases were tested and compared.

In terms of energy yield capabilities, the results showed that:

- the intrinsic capability of H-Darrieus rotors to take advantage from moderately skewed flows can lead to notable increases of their producibility: for an attended skew angle of 20°, the increase of energy conversion with respect to the aligned flow case was estimated in approximately +40%. Beyond the limit value of $\gamma = 30°$, a decrease of energy conversion was instead noticed (approximately −9%);
- a rotor specifically optimized to work at a prescribed specific skew angle can provide additional benefits in terms of energy production (up to +20% at $\gamma = 20°$) with respect to the rotor originally optimized for aligned flow (which, however, also performs better in skewed flow).

Focusing instead on the design parameters, the optimal design solution in skewed flow was shown to be fairly independent on the average wind speed. In these conditions, the optimal design is in fact defined by the maximization of the skewed flow effects. In particular, comparing the optimum design configurations with their homologous ones in aligned slow, it can be noticed that:

- larger diameters are preferable, since they maximize the increase of the swept area, i.e. the wind flow "captured" by the turbine;

- the upper limit in the maximization of the diameter is imposed by a balance with the blade efficiency loss due to tip effects (function of the blade Aspect Ratio), which defines the minimum blade height;
- for a given average wind speed, the larger diameters increase the peripheral speed and then the blade Reynolds numbers with respect to the same cases in aligned flow. As a result, the chord can be generally shorter, mitigating the problems in terms of reduced blade AR;
- as a consequence of the above, the optimal rotors in skewed flow are generally characterized by lower solidities with respect to the aligned-flow cases.

Finally, it is worth pointing out that aforementioned results can be thought of particular interest for future applications of H-Darrieus turbines in the urban environment, with particular reference to rooftop installations, in which the attended skew angle can be reasonably considered constant and a priori estimated upon analysis of the surrounding environment. In these conditions, a wind turbine specifically conceived to work under skewed flow can provide substantial increases of the energy yield, becoming then competitive with horizontal-axis wind turbines, which instead suffer from a constant performance decrease in skewed/yawed flow conditions.

Acknowledgements Thanks are due to Prof. Ennio Antonio Carnevale of the University of Florence for supporting this research activity.

References

1. GWEC: Global Wind Report 2016 (2016). http://www.gwec.net/wp-content/uploads/vip/GWEC-Global-Wind-Report_2016.pdf
2. WWEA: 2016 Small Wind World Report. Technical report, Washington DC, USA (2016)
3. Colak, I., Fulli, G., Bayhan, S., Chondrogiannis, S., Demirbas, S.: Critical aspects of wind energy systems in smart grid applications. Renew. Sustain. Energy Rev. **52**, 155–171 (2015)
4. Kaldellis, J.K.: Optimum autonomous wind–power system sizing for remote consumers, using long-term wind speed data. Appl. Energy **71**, 215–218 (2002)
5. Proietti, S., Sdringola, P., Castellani, F., Garinei, A., Astolfi, D., Piccioni, E., Desideri, U., Vuillermoz, E.: On the possible wind energy contribution for feeding a high altitude smart mini grid. Energy Proc. **75**, 1072–1079 (2015)
6. AWEA: Small & Community Wind. Technical report (2016)
7. Mertens, S.: The energy yield of roof mounted wind turbines. Wind Eng. **27**(6), 507–517 (2003)
8. Dayan, E.: Wind energy in buildings: power generation from wind in the urban environment—where it is needed most. Refocus **72**(2), 33–38 (2006)
9. Balduzzi, F., Bianchini, A., Ferrari, L.: Microeolic turbines in the built environment: influence of the installation site on the potential energy yield. Renew. Energy **45**, 163–174 (2012)
10. Mertens, S.: Wind Energy in the Built Environment. Multi-Science, Brentwood, UK (2006)

11. Paraschivoiu, I.: Wind Turbine Design with Emphasis on Darrieus Concept. Polytechnic International Press, Canada (2002)
12. Aslam Bhutta, M.M., Hayat, N., Farooq, A.U., Ali, Z., Jamil, Sh.R., Hussain, Z.: Vertical axis wind turbine—a review of various configurations and design techniques. Renew. Sustain. Energy Rev. **16**(4), 1926–1939 (2012)
13. Sharpe, T., Proven, G.: Crossflex: concept and early development of a true building integrated wind turbine. Energy Build. **42**, 2365–2375 (2010)
14. Mertens, S., van Kuik, G., van Bussel, G.: Performance of an H-Darrieus in the skewed flow on a roof. J. Sol. Energy Eng. **125**, 433–440 (2003)
15. Bianchini, A., Ferrari, L., Magnani, S.: On the effects of a skewed flow on the performance of a three-bladed H-Darrieus turbine: experimental and theoretical analyses. In: Proceedings of the International Conference on Applied Energy (ICAE) 2012, Suzhou (China), July, pp. 5–8 (2012)
16. Simão Ferreira, C.J., van Bussel, G., van Kuik, G.: An analytical method to predict the variation in performance of a H-Darrieus in skewed flow and its experimental validation. In: Proceedings of the European Wind Energy Conference, Athens, Greece (2006)
17. Bianchini, A., Ferrara, G., Ferrari, L., Magnani, S.: An improved model for the performance estimation of an H-Darrieus wind turbine in skewed flow. Wind Eng. **36**(6), 667–686 (2012)
18. Bianchi, S., Bianchini, A., Ferrara, G., Ferrari, L.: Small wind turbines in the built environment: influence of flow inclination on the potential energy yield. J. Turbomach. **136**(4), 1–9 (2014)
19. Balduzzi, F., Bianchini, A., Carnevale, E.A., Ferrari, L., Magnani, S.: Feasibility analysis of a Darrieus vertical-axis wind turbine installation in the rooftop of a building. Appl. Energy **97**, 921–929 (2012)
20. Borg, M., Shires, A., Collu, M.: Offshore floating vertical axis wind turbines, dynamics modelling state of the art. Part I: aerodynamics. Renew. Sustain. Energy Rev. **39**, 1214–1225 (2014)
21. Bianchini, A., Ferrari, L., Magnani, S.: Energy-yield-based optimization of an H-Darrieus wind turbine. In: Proceedings of the ASME Turbo Expo 2012, Copenhagen, Denmark, June, pp. 11–15 (2012)
22. Beller, C.: Urban wind energy—state of the art 2009, Risø Laboratory—DTU, Roskilde, Denmark, Technical report Risø-R-1668 (EN) (2009)
23. Manwell, J.F., McGowan, J.G., Rogers A.L: Wind Energy Explained, 2nd edn. Wiley, UK (2009)
24. Banks, D., Cochran, B., Denoon, R., Wood, G.: Harvesting wind power from tall buildings. In: Proceedings of the CTBUH 8th World Congress, Dubai (UAE) (2008); In: Chicago: Council on Tall Buildings and Urban Habitat, pp. 320–327 (2008)
25. Bianchini, A., Ferrara, G., Ferrari, L.: Design guidelines for H-Darrieus wind turbines: optimization of the annual energy yield. Energy Convers. Manag. **89**, 690–707 (2015)
26. Islam, M.R., Saidur, R., Rahim, N.A.: Assessment of wind energy potentiality at Kudat and Labuan, Malaysia using Weibull distribution function. Energy **36**(2), 985–992 (2011)
27. Ferrari, L., Bianchini, A.: Critical aspects in the design of a small-size Darrieus wind turbine. In: Proceedings of the World Renewable Energy Congress (WREC) XI, Abu Dhabi, UAE, Sept, pp. 25–30 (2010)
28. Bianchini, A., Balduzzi, F., Rainbird, J., Peiro, J., Graham, J.M.R., Ferrara, G., Ferrari, L.: On the influence of virtual camber effect on airfoil polars for use in simulations of Darrieus wind turbines. Energy Convers. Manag. **106**(December 2015), 373–384 (2015)
29. Bianchini, A., Balduzzi, F., Rainbird, J., Peiro, J., Graham, J.M.R., Ferrara, G., Ferrari, L.: An experimental and numerical assessment of airfoil polars for use in Darrieus wind turbines. Part 1—Flow Curvature Effects. J. Eng. Gas Turbines Power **138**(3), 1–10 (2015)

30. Bianchini, A.: Performance analysis and optimization of a Darrieus VAWT. PhD Thesis, School of Energy Engineering and Innovative Industrial Technologies, University of Florence, Italy (2011)
31. Bianchini, A., Cangioli, F., Papini, S., Rindi, A., Carnevale, E.A., Ferrari, L.: Structural analysis of a small H-Darrieus wind turbine using beam models: development and assessment. J. Turbomach. **137**(1), 1–11 (2015)
32. Paraschivoiu, I., Delclaux, F.: Double multiple streamtube model with recent improvements. J. Energy **7**(3), 250–255 (1983)
33. Bianchini, A., Balduzzi, F., Ferrara, G., Ferrari, L.: Critical analysis of dynamic stall models in low-order simulation models for vertical-axis wind turbines. Energy Proc. **101**, 488–495 (2016)
34. Jones, R.T., Cohen, D.: Aerodynamics of wings at high speeds. In: Donovan, A.F., Lawrence, H.R. (eds.) Aerodynamic Components of Aircraft at High Speeds, High Speed Aerodynamics and Jet Propulsion, vol. VII. Princeton University Press, Princeton, USA (1957)
35. Bianchini, A., Balduzzi, F., Bachant, P., Ferrara, G., Ferrari, L.: Effectiveness of two-dimensional CFD simulations for Darrieus VAWTs: a combined numerical and experimental assessment. J. Energy Convers. Manag. **136**, 318–328 (2017)

Numerical and Experimental Comparison of Performance of Two Stage and Helical Savonius Wind Turbines

L. Brito Kothe and A. Prisco Petry

Abstract This paper presents a numerical and experimental comparison of vertical axis wind turbine involving two stage and helical Savonius rotors. The experimental study is conducted in the Aerodynamic Tunnel Professor Debi Pada Sadhu at the Fluid Mechanics Laboratory of the UFRGS. The rotors are manufactured by 3D prototyping technique. The numerical simulations are performed using the Finite Volumes Method. The dynamic torque coefficient, power coefficient, and an aerodynamic analysis of the two turbines are compared for tip speed ratios (λ) between 0.2 and 0.8. The torque and power coefficients of the helical turbine are higher than the two stage turbine for most cases. The helical turbine shows less torque variation along each rotation when compared with the two stage turbine. The differences between the numerical and experimental values obtained are between 6.17 and 13.3% for the two stage turbine and between 2.34 and 13.11% for the helical turbine.

Keywords Vertical axis wind turbines · Numerical study · Experimental study

Nomenclature

A	Area, m^2
c	Chord, m
C_P	Power coefficient
C_T	Torque coefficient
d_h	Hole diameter, m
d_b	Bucket diameter, m
e	Thickness, m
e_{pe}	Thickness of the buckets, m
F	Force, N

L. Brito Kothe (✉) · A. Prisco Petry
Mechanical Engineering Department, Federal University of Rio Grande do Sul, Sarmento Leite Street, 425, Porto Alegre 90050-170, Brazil
e-mail: leonardo.kothe@ufrgs.br

L. Battisti and M. Ricci (eds.), *Wind Energy Exploitation in Urban Environment*, Green Energy and Technology, https://doi.org/10.1007/978-3-319-74944-0_13

g	Gravity, m/s^2
h_1	Turbine stage height, m
h_2	Thickness of the intermediate plate, m
k	Turbulent kinetic energy, J/kg
$m_{BALANCE}$	Balance mass, kg
m_{LOAD}	Load mass, kg
P	Power, W
r	Rotor radius, m
r_t	Total radius, m
r_{SHAFT}	Shaft radius, m
Re	Reynolds number
s	Overlap, m
t_{STRING}	String thickness, m
T	Moment, N·m
V_o	Undisturbed air flow velocity, m/s
u	Velocity flow, m/s
u'	Velocity fluctuation, m/s
$\bar{u}$	Average velocity, m/s
Δt	Time step, s
$\Delta\theta$	Angular step, °
μ	Air dynamic viscosity, kg/(m·s)
ν	Air kinematic viscosity, m^2/s
θ	Angular position, °
ρ	Atmospheric air density, kg/m^3
λ	Tip speed ratio of the rotor
ω	Specific turbulence dissipation rate, s^{-1}
ω_o	Angular velocity, rad/s
τ	Reynolds stress

1 Introduction

Responsible to transform the kinetic energy of wind in mechanical energy, the wind turbines has become a major means of alternative and renewable energy production in order to reduce environmental pollution and exhaustion of non-renewable sources. These types of turbines can be classified into horizontal and vertical axis. The purpose of this article is to study a vertical axis wind turbine, which it is emphasized the Savonius type, presenting a simple construction when compared with other types [1].

The Savonius rotor type, invention patented by Sigurd J. Savonius in 1920, presents a vertical axis, where its original formatting had the blades in a S-shaped form placed along the axis which the rotation occurs mainly due to the drag force provided by the wind in its concave and convex buckets [2]. Initially, this type of rotor is commonly used for pumping stations and driving force due to its high torque developed and low angular velocity. However, changes such as an increase in its aspect ratio can be made, providing higher velocity and increasing generation of power [3–5], especially when the cost is more important than efficiency, also could be installed in combination with other types of turbines or other types of generation, such as photovoltaic cell. This type of small turbines can be used in places such as homes, commercial farms and buildings which can cover part or all of the energy required. Another advantage of this type of turbine is the suitability to operate in urban centers due to the ability to take advantage of the winds with large and fast direction changes [4, 6].

In the literature it is possible to find experimental and numerical studies seeking to represent and optimize the main features of this type of turbine. For conventional Savonius rotors, studies were performed to analyze performance coefficients, varying characteristics such numbers of stages, aspect ratios and overlap ratio. Fujisawa [3] conducted an experimental study with a comparison between different overlap ratio for a rotor with and without rotation. For the static torque, the performance was improved by increasing of overlap ratio, especially on the return bucket due to the effect of pressure recovery by flow through this overlap. For the rotation rotor the peak values of the torque and power coefficients was achieved for an overlap ratio of 0.15. Saha et al. [4] conducted tests in a wind tunnel to get results on the aerodynamic performance of rotor with one, two and three stages. The authors mixed the format of the buckets between semicircular and twist, the number of buckets and the use of valves on the buckets. The main observations of the authors is that the optimal number of buckets found has two, both for one, two or three stages. Two stage rotors had higher power coefficients with respect to one and three stage rotors.

In recent years, computational solutions were used by authors such as [7, 8]. Menet and Bourabaa [7] checked the behavior of a rotor for Reynolds values between 1×10^5 and 5×10^5. For this range of values, there is an increase in performance coefficients for almost the entire range, with the highest values obtained for the largest Reynolds values. Akwa et al. [8] simulated Savonius rotors with different overlap ratio using the Reynolds Averaged Navier-Stokes equations methodology (RANS) and turbulence model *SST* k-ω. Among the analyzed settings, the best performance was obtained for 0.15 which a power coefficient of 0.3161 for a tip speed ratio (λ) of 1.25. According to the authors, a space between the buckets permits the passage of air from the forward to return bucket, increasing the pressure of the back side of concave bucket and reducing drag force on this part of the rotor.

In view to improve the performance of Savonius turbines, seeking greater uniformity in torque, some authors [6, 9–12] proposed the study of a helical Savonius turbine with a twist along its vertical axis. According to [6], the 180° twist buckets provides more uniformity in the coefficient of dynamic torque of the turbine and less oscillatory behavior of the torque. Kamoji et al. [9] presented results indicating that the helical rotors have positive static torque coefficient for every rotor angle.

According to published studies, it was possible to verify that the performance of Savonius rotors can be improved. It was observed also discrepancies between numerical and experimental results due the differences found between the used methodologies. In this paper was presented a comparative study between the numerical

Table 1 Chosen parameters for the turbines

Parameters	Two stage dimensions (mm)	Helical dimensions (mm)
D	95.0	95.0
H	380.0	380.0
d_f	4.5	4.5
d_{pe}	104.5	104.5
e_{pe}	7.5	7.5
e	5.0	5.0
h_1	187.5	–
h_2	5.0	–
c	51.35	51.35
s	7.7	7.7

and experimental analysis on the performance of a helical and two stage Savonius turbines.

2 Methodology

2.1 Wind Turbine Models

For the comparison be performed numerically and experimentally it is necessary to maintain the same characteristics, such as height, diameter, overlap ratio and size of the end plates. The both turbines are first designed by a 3D CAD software and then is manufactured through 3D prototyping. The numerical simulations are performed with the same geometry created by 3D CAD.

Figure 1 shown a comparison of the dimensions of Savonius turbines. The height is 0.380 m, the rotor diameter is 0.095 m and the thickness of the buckets is 0.005 m, so that the maximum frontal area projected by the rotor is 0.0361 m^2, not requiring blockage effect correction as the cross-sectional area of the test section has 1 m^2. The buckets have an overlap of 15%. According to Akwa [13], overlaps with values between 10 and 30% of the chord give good results for the average power ratio in operation. In Table 1 are described all the parameters necessary for manufacturing.

The diameter of the end plates is 0.1045 m, and their thickness is 0.0075 m. According to Akwa [13], end plates improves the aerodynamic performance of the rotor, because its presence prevents air leakage in the concave part of the buckets to the external flow, while maintaining the differential pressure in the concave and convex parts of bucket.

The aspect ratio for these turbines is 4.0. According to Alexander and Holownia [14], high aspect ratios should improve efficiency.

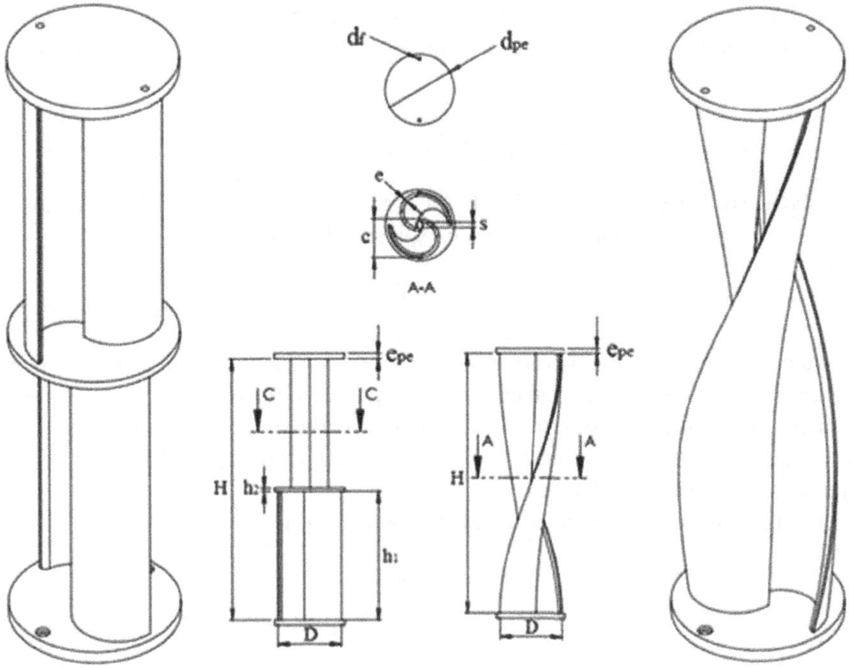

Fig. 1 Two stage (left) and helical (right) Savonius rotor

2.2 Numerical Methodology

Numerical simulations of the Savonius rotors are performed in the commercial software Fluent/Ansys using Finite Volumes Method. Finite Volumes Method uses the integral form of the conservation equations as its starting point [15]. The solution domain is divided into a finite number of adjacent control volumes, and the conservation equations are applied to each control volume. An algebraic equation is obtained for each volume control that is solved iteratively for each volume [16]. The mass conservation equations and linear momentum are solved by the Reynolds Averaged Navier-Stokes (RANS) equations.

Reynolds decomposition consisting in decomposes a scalar or vector variable in a medium temporal part and another as floating part, as Eq. 1 [17]. It is possible to make a replacement in the principle of mass conservation, Eq. 2:

$$u(t) = \bar{u} + \bar{f}(\vec{x}, t) \tag{1}$$

$$\frac{\partial \bar{u}_i}{\partial x_i} = 0 \tag{2}$$

For a given notation, with Reynolds decomposition for velocity and pressure by the momentum principle, the balance equation of momentum is obtained as Eq. 3:

$$\frac{\partial \bar{u}_i}{\partial t} + \bar{u}_j \frac{\partial \bar{u}_i}{\partial x_j} = -\frac{1}{\rho} \frac{\partial \bar{p}}{\partial x_i} + \nu \frac{\partial^2 \bar{u}_i}{\partial x_j \partial x_j} - \frac{\partial}{\partial x_j} \overline{u'_i u'_j} \tag{3}$$

The tensor correlations described by $(u'_i u'_j)$, composes the Reynolds stress tensor, which is a symmetric tensor and can be expressed by (Eq. 4):

$$-\rho \overline{u'_i u'_j} \tag{4}$$

According to [17], the Reynolds stress represents the reciprocal of fluids due to the presence of fluctuations, and not due to the proper tensor. Its increase in Reynolds equation generates a closure problem caused when there are more unknowns than equations. To solve this problem, turbulence models are used. For this paper is used the turbulence model *SST k-ω*.

For the time step value in each case it is used the Eq. 5:

$$\Delta t = \frac{\Delta \theta \pi}{180^\circ \omega_o} \tag{5}$$

2.2.1 Computational Domain

The simulations with stationary turbine, which are used for obtaining the evaluation of mesh quality, are steady and three-dimensional. The simulations with rotating turbine are transient, three-dimensional, with a turbulent length scale of 0.07 m and using a turbulence intensity of 2%. The simulated tip speed ratios are 0.2, 0.35, 0.5, 0.65 and 0.8. For each case is used three complete turns of the turbine and an angular displacement (Δθ) of 2.5°.

The computational domain and the boundary conditions are presented in Fig. 2. The length of the computational domain is 6 m and its cross area is 1 m^2. The turbine is located downstream 2 m of the inlet region. The computational domain is based on the aerodynamic tunnel dimensions used for the experimental tests. The interface region includes the simulated turbine located at the center. Thus it is possible to impose an angular velocity in the turbine without changing the prescribed velocity inlet. The convergence criterions used in the solutions are 10^{-4} for the steady simulations and 10^{-5} for transient simulations. The boundary condition at the inlet region has a velocity of 7.6 m/s. For the outlet region was considered pressure outlet.

The Reynolds number is calculated from Eq. (6):

$$\mathrm{Re} = \frac{\rho D V_o}{\mu} \tag{6}$$

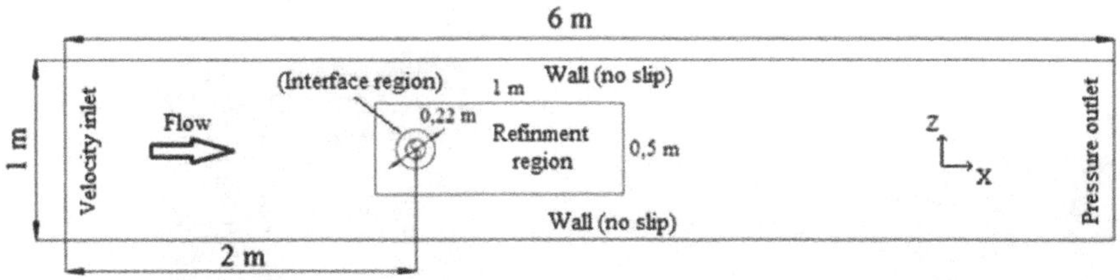

Fig. 2 Computational domain

For D being the rotor diameter and V_o the undisturbed flow velocity, providing a Reynolds number of 49,640 for a temperature of 15 °C.

2.3 Experimental Methodology

Experimental tests are performed in the open circuit aerodynamic tunnel at the Fluid Mechanics Laboratory of the UFRGS.

With the support of the Information Technology Center Renato Archer, the models are produced in polyamide (nylon), a material that has good strength, by a laser printer—Fig. 1 as described in Sect. 2.1. Due to small imperfections in the printed form, it is performed a post process with the application of rubbing compound, sanding and painting process. Figure 3a shows the turbine before and after the finishing process. Figure 3b shows the support for fixing the turbine in the aerodynamic tunnel. The support consists in a vertical wooden pole with two horizontal bars, one upper and one lower, which are cable-stayed at the top and bottom of the tunnel in order to reduce vibrations due to the flow. Each bar has a rolling bearing, a shaft attached to the bearing and an attached disk in the shaft end. Between these two discs is that the turbine is located. To reduce the flow interference in the turbine, the support bracket is attached on the side wall of the aerodynamic tunnel.

The tests are performed for an inlet velocity of 7.6 m/s at the voltage source inverter set to 40 Hz. The inlet velocity is checked by a Pitot tube situated in the test section upstream of 0.5 m. The turbine rotation is measured with a digital photo-tachometer, model MDT 2244B. This instrument operates in a range of 2.5–99,999 rpm and has an accuracy of $\pm(0.05\% + 1\text{D})$.

Due to the need for more results, it has been proposed to manufacture a simplified dynamic torque wrench similar to that used by [5, 11]. For the assembly of the system is necessary: a string, a digital balance, a pulley and different portions of weights. The operating principle is to create a contact between the string and the axis of the turbine. Attached to one end of the string is a digital balance, with 5 g resolution. On the other side, the string pass by a pulley and go down to an outside region of the aerodynamic tunnel where this tip will be placed loads to function as a brake. Each load is weighed before the test in a digital precision balance, with 0.1 g resolution.

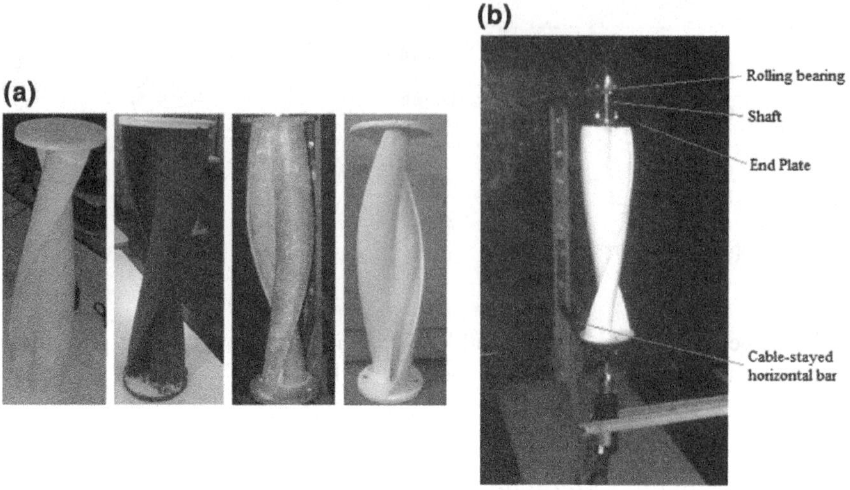

Fig. 3 Savonius helical turbine model: **a** printed model and **b** support

With the rotor in operation, the rotation of the rotor will pull a load on the digital balance, and thus, the resultant force (F) on the shaft can be calculated by Eq. (7).

$$F = (m_{BALANCE} - m_{LOAD})\, g \tag{7}$$

where $m_{BALANCE}$ is the mass measured in the balance; m_{LOAD} is the mass of the load hanging on the string and g is the gravity. To determine the dynamic torque (T), Eq. (8), a multiplication of the resultant force by the radius (r) of the shaft is made, the rotor radius is given by $r = r_{SHAFT} + t_{STRING}$ where r_{SHAFT} is the radius of the shaft with 0.005 m and t_{STRING} is the string thickness which has 0.0022 m.

$$T = F \cdot r \tag{8}$$

At the moment when the turbine is rotating it is necessary to measure the angular velocity (ω_o) to determine the tip speed ratio and the estimated power as described in Eq. (9):

$$P = T \cdot \omega_o \tag{9}$$

Torque coefficient (C_T) and power coefficient (C_P) are determined by the Eq. (10) and Eq. (11), respectively.

$$C_T = \frac{T}{\frac{1}{2}\rho A V_o^2 r} \tag{10}$$

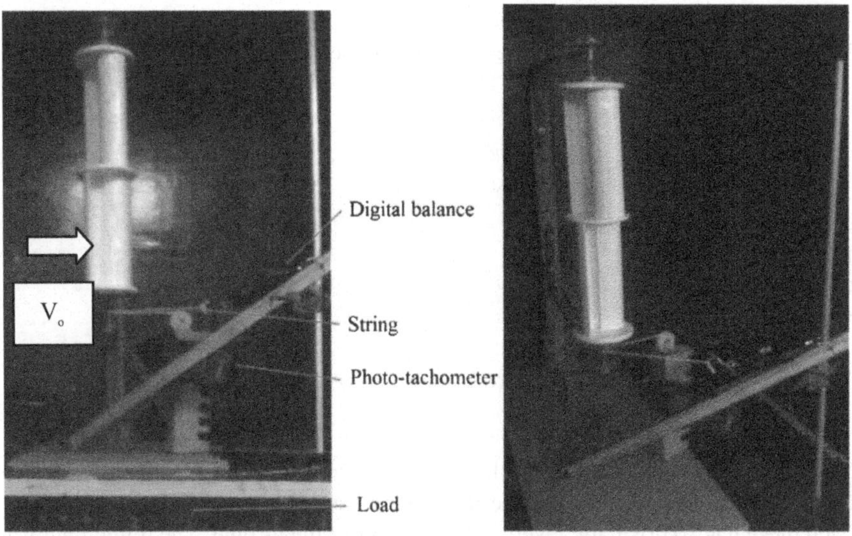

Fig. 4 Dynamic torque wrench scheme

$$C_P = \frac{T\omega}{\frac{1}{2}\rho A V_o^3} \tag{11}$$

In Fig. 4 a schematic picture of the photo tachometer and the system with the digital balance, string, pulley and load is displayed.

A more accurate method for estimating the uncertainty of results was presented by [18]. His method is based on a detailed specification of the uncertainties of the different primary experimental measurements. Assuming a series of measurements is made and the uncertainty in each measurement can be expressed in the same form, it is then possible to use them to calculate the uncertainty of the experiments [19].

For the results, the measurement uncertainties are considered and the used equations are presented in [20].

2.4 Mesh Quality Evaluation

The mesh refinement of this paper is accomplished by the *GCI* method (Grid Convergence Index). According to Roache [21], the *GCI* is based on the generalized Richardson extrapolation involving comparison of discrete solutions in two different spacing meshes.

The application of *GCI* method is performed for three tetrahedral meshes with different refinements. Each mesh has an interface region, which has tetrahedral volumes, and a region representing the aerodynamic tunnel domain. The aerodynamic

Table 2 Mesh refinement

Mesh	Number of volumes in the rotor region	Total number of volumes
M1	6,068,584	9,750,624
M2	3,491,682	7,173,722
M3	2,244,807	5,926,847

Table 3 Results of static torque coefficient for each simulated mesh

Mesh	Static torque coefficient
M1	0.3744
M2	0.3948
M3	0.4248

tunnel domain is invariant to the three meshes and composed with 3,682,040 volumes. The area near the turbine interface region has different refinement, as Table 2. Each mesh has 10 prismatic layers in the contour of the turbine. The refinement was performed for the mesh of the helical turbine.

The results of the static torque coefficients for the three simulated meshes are shown in Table 3.

With the results of the static torque coefficient of each mesh (Table 3) is possible to apply the *GCI* method. The found value for the estimation of exact solution of the torque coefficient for the finest meshes ($M1$ and $M2$) is 0.3656, the convergence rates for the meshes (GCI_{12} and GCI_{23}) are 3.407% and 9.691%, respectively. The value of asymptotic convergence is 0.948. The application of the method indicates that for values close to 1, the solution is within the range of asymptotic convergence.

Based on the results it is possible to conclude that the solution is within the range of asymptotic convergence and the rotor static torque ratio is estimated to be 0.3656%. The mesh with the bigger refinement, with 9,750,131 volumes, is chosen for the paper continuity due to the best results. The same mesh parameters were used to create the mesh of the two stage turbine, obtaining a mesh with 9,760,131 volumes.

3 Results and Discussion

This section presents comparisons of performance of the helical and two stage turbines. Comparisons between numerical and experimental results are also shown. Tests are performed to λ values of 0.2, 0.35, 0.5, 0.65 and 0.8 for numerical and experimental cases, where the values of λ vary with the angular velocity of the turbine. At the end, it is presented a curve of torque for different angles of rotation of the turbines and similar results found in the literature.

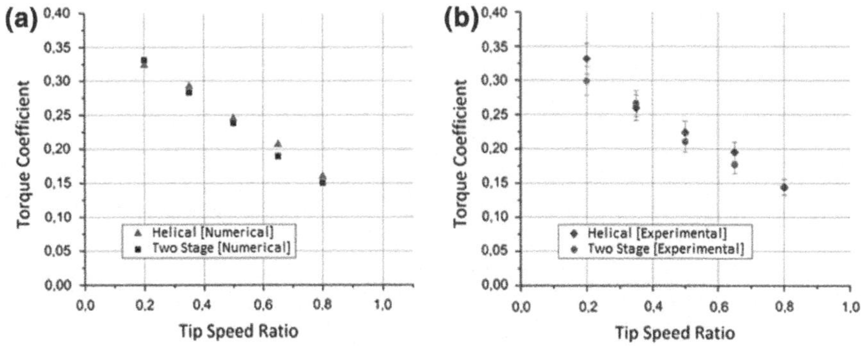

Fig. 5 Comparison of torque coefficient results for: **a** numerical cases and **b** experimental cases

3.1 Comparison of Performance Coefficients

For the measurements are considered the pressure and temperature conditions of the laboratory at the time of the tests, being 750 mm Hg and 24 °C, respectively. In a free condition, for a flow velocity of 7.6 m/s is obtained a rotation of 160.32 rad/s for the helical turbine and 130.37 rad/s for the two stage turbine. Because of this, for the helical and two stage turbines it is possible to conduct measurement only for lower λ values than 1.0 and 0.8, respectively, by issues of sensitivity of a measuring system.

In a comparison of torque coefficients, graphics for different tip speed ratios are presented. In Fig. 5 is shown a similarity between the values. The helical turbine showed higher values in general, however, due to uncertainties in experimental measurements, this difference can vary up to ±0.02292 for the helical turbine and ±0.0208 for the two stage turbine in the experimental results. The maximum relative difference in the numerical comparison is 9.3% for a tip speed ratio of 0.65, as in the experimental comparison, the difference is 9.9% for a λ value of 0.2.

In a comparison of power coefficients, Fig. 6 shows a greater difference in the maximum coefficient obtained for the tip speed ratio of 0.65. The relative difference obtained for the both cases are the same that is obtained for de C_T values. In the experimental comparison, due to uncertainties, the difference of the results can vary up to ±0.0101 for the helical turbine and ±0.00938 for the two stage turbine. In general, the helical turbine showed higher values.

3.2 Experimental and Numerical Comparison

The numerical and experimental power coefficients of the turbines are presented in Fig. 7 for a comparison. For the helical turbine (a), it can be seen that the C_P values are similar for both cases, with a larger relative difference of 13.11% for a λ of 0.35 is

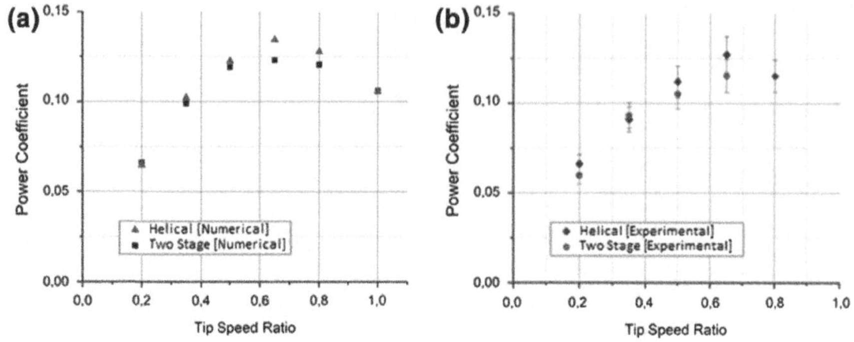

Fig. 6 Comparison of power coefficient results for: **a** numerical cases and **b** experimental cases

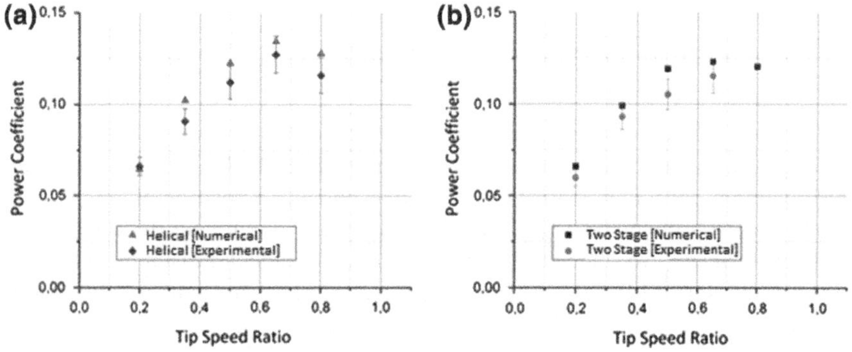

Fig. 7 Numerical and experimental comparison of power coefficients: **a** helical and **b** two stage

found. The lowest difference found is 2.34% for a λ of 0.2. This percentage difference is the difference between the values found in the numerical and experimental tests divided by the value found in the torque coefficient in the experimental test. For the two stage turbine (b), it is may be seen that is a maximum of C_P for the tip speed ratio of 0.65 where the value grows to this value, and then decreases again. The relative differences between the values obtained are between 6.17 and 13.3%.

3.3 Comparison of a Torque Curve in a Revolution

In this section is compared the torque simulated numerically for a complete turn of the turbine. For stability, the results are taken for the third round of the turbines. Figure 8 shows a different behavior between helical and two stage turbines. In the helical turbine it is possible to visualize a smaller oscillation of the torque along the turn. As for the two stage turbine occurs a greater torque oscillation, which is not positive for power generation. In addition, the helical turbine has two largest peaks

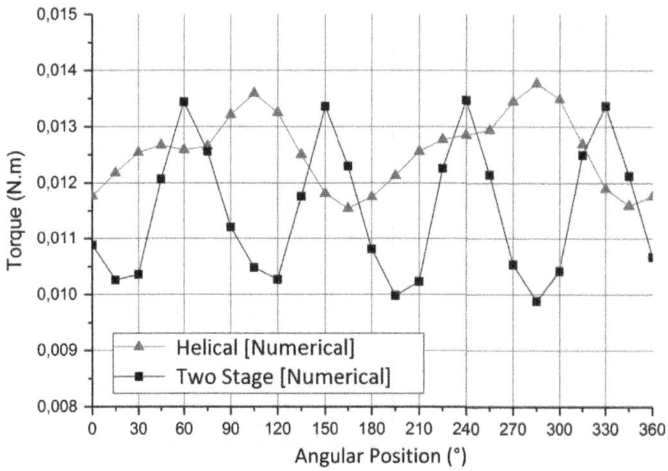

Fig. 8 Comparison of torque for different rotation angles for the helical and two stage turbines

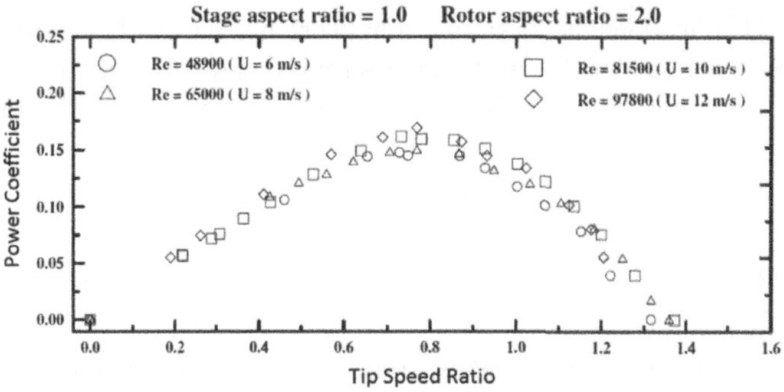

Fig. 9 Power coefficient curve of a two stage turbine (adapted by [9])

in the torque fluctuation; however, the two stage turbine has four peaks of torque. The average torque for a turn is 0.01255 Nm for the helical turbine and 0.01149 Nm for the two stage turbine, showing superiority for the helical turbine by 8.47%.

3.3.1 Comparison with Others Savonius Rotors

The results obtained for the turbines are compared with the results already reported in the literature. Kamoji [9] conducted an experimental study on Savonius rotor with one, two and three stages. The two stages rotor showed an aspect ratio of 2.0 and an overlap of 15%. From the experimental tests, it was possible to create a power coefficient graph (Fig. 9) for the two stages turbine for different Reynolds numbers ranging from 48,900 to 97,800.

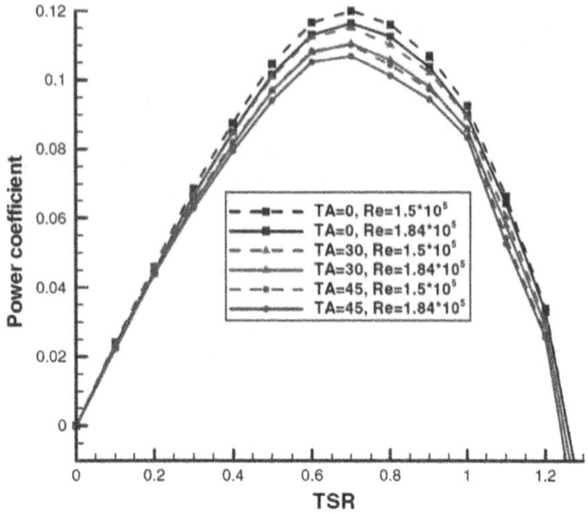

Fig. 10 Power coefficient curve of a helical turbine (adapted by [10])

In Jeon [10], it is obtained a similar trend behavior for a turbine with a twist of 180° and an aspect ratio of 2.0. In the study, turbines are tested in different conditions and for different flow velocities. In Fig. 10 is showed the power coefficient behavior for different values of tip speed ratio, presenting maximum values near of 0.65.

4 Conclusions

This paper presents a numerical and experimental comparative analysis of the performance of helical and two stage Savonius turbines, manufactured by 3D prototyping method. The 3D prototyping technique is used because of the ease to create models of great complexity, an important characteristic for comparative studies due to the necessity to generate models while maintaining the same dimensions and material.

In a comparison between the helical and two stage performances it is possible to note a similarity of behavior in the coefficients. The performance results of helical turbine are superior, however, due to measurement uncertainties, the results may be closer. The helical turbine also has a smaller oscillation of the torque along the turn. Moreover, the two stage turbine was easier to manufacture due to its direct formatting of buckets.

In a comparative between numerical and experimental results, for the helical turbine, in both cases the maximum tip speed ratio value for the power coefficient is given for 0.65. The maximum value of torque coefficient is also approximate in the numerical and experimental studies, where the maximum value is obtained for a λ of 0.2. The average differences between the results are maintained between 2.34 and 12.5%. For the two stage turbine, the maximum tip speed ratio values are also 0.65

for the power coefficient and 0.2 for the torque coefficient. The average differences of the results are in order of 3.8–13.4%.

The used method for assessing the mesh quality is significant in chosen the most appropriate mesh and quantification of the intrinsic error. The torque wrench showed similar results to those found numerically.

The obtained results are important to increase characteristics of performance of Savonius wind turbines in the literature, adding numerical and experimental values, and also showing the importance of increasing the numerical methodology in this type of research. For future experiments, hybrid forms and open field simulations can be performed.

Acknowledgements The authors would like to acknowledge Capes and CNPq for the financial support of this research.

References

1. Burton, T., Jenkins, N., Sharpe, D., Bossanyi, E.: Wind Energy Handbook, 2nd edn. Wiley (2011)
2. Savonius, S.J.: Wind Rotor—Patent Number 1,766,765, United States Patent Office (1930)
3. Fujisawa, N.: On the torque mechanism on Savonius rotors. J. Wind Eng. Ind. Aerodyn. **40**(3), 277–292 (1992)
4. Saha, U.K., Thotla, S., Maity, D.: Optimum design configuration of Savonius rotor though wind tunnel experiments. J. Wind Eng. Ind. Aerodyn. **96**, 1359–1375 (2008)
5. Kamoji, M.A., Kedare, S.B., Prabhu, S.V.: Experimental investigations on single stage modified Savonius rotor. Appl. Energy **86**, 1064–1073 (2009)
6. Oliveira, C.P., Petry, A.P.: Numerical study of a helical Savonius wind turbine. In: 15th Brazilian Congress of Thermal Sciences and Engineering, ENCIT 2014, Belém do Pará (2014)
7. Menet, J.L., Bourabaa, N.: Increase in the Savonius rotors efficiency via a parametric investigation. In: European Wind Energy Conference, London (2004)
8. Akwa, J.V., Silva Jr., G.A., Petry, A.P.: Discussion on the verification of the overlap ratio influence on performance coefficients of a Savonius wind rotor using computational fluid dynamics. Renew. Energy **38**, 141–149 (2012)
9. Kamoji, M.A., Kedare, S.B., Prabhu, S.V.: Experimental investigations on single stage, two stage and three stage conventional Savonius rotor. Int. J. Energy Res. **32**, 877–895 (2008)
10. Jeon, K.S., Jeong, J.I., Pan, J.-K., Ryu, K.-W.: Effects of the end plates with various shapes and sizes on helical Savonius wind turbines. Renew. Energy **79**, 167–176 (2015)
11. Anbarsooz, M.: Aerodynamic performance of helical Savonius wind rotors with 30° and 45° twist angles: experimental and numerical studies. J. Power Energy, 1–10 (2016)
12. Zhao, Z., Zheng, Y., Xu, X., Lui, W., Zhou, D.: Optimum design configuration of helical Savonius rotor via numerical study. In: ASME 2009 Fluids Engineering Division Summer Meeting, Vail, Colorado, USA, 2–6 Aug 2009
13. Akwa, J.V., Vielmo, H.A., Petry, A.P.: A review on the performance of Savonius wind turbines. Renew. Sustain. Energy Rev., 3054–3064 (2012)
14. Alexander, A.J., Holownia, B.P.: Wind tunnel tests on a Savonius rotor. J. Ind. Aerodyn. **3**(4), 343–351 (1978)
15. Ferziger, J.H., Peric, M.: Computational Methods for Fluid Dynamics, 3rd edn. Springer, Berlin, Heidelberg, NY, Barcelona, Hong Kong, London, Milan, Paris, Tokyo (2002)
16. Patankar, S.V.: Numerical heat transfer and fluid flows. McGraw-Hill, NY, USA (1980)
17. Wilcox, D.C.: Turbulence Modeling for CFD. DCW Industries, Inc. (1998)

18. Kline, S.J., McClintock, F.A.: Describing the uncertainties in single sample experiments. Mech. Eng., 3–8 (1953)
19. Holman, J.P.: Experimental Methods for Engineers, 8th edn. McGraw-Hill, NY, USA (2012)
20. Kothe, L.B., Estudo comparativo experimental e numérico sobre o desempenho de turbinas savonius helicoidal e de duplo-estágio. Master thesis, PROMEC-UFRGS, Porto Alegre, Brazil (2016)
21. Roache, P.J.: Perspective: a method for uniform reporting of grid refinement studies. J. Fluids Eng. **116**, 405–413 (1994)

MPPT with Current Control for a PMSG Small Wind Turbine in a Grid-Connected DC Microgrid

D. Zammit, C. Spiteri Staines, A. Micallef and M. Apap

Abstract This paper presents a Maximum Power Point Tracking (MPPT) system for a small wind turbine (SWT) connected to a DC Microgrid under grid-connection conditions. The system consists of a Permanent Magnet Synchronous Generator (PMSG) driven by a SWT which is interfaced to the DC microgrid through a rectification stage and boost converter. The proposed MPPT system is based on the relationship between the DC link power and voltage, which are used to obtain the required inductor current in the Boost converter to provide maximum power output at all wind speeds. The MPPT system, the Boost converter and current control are explained in detail. Simulation of the system operating under varying wind conditions is presented, showing the performance of the developed MPPT algorithm.

Keywords MPPT · PMSG · Boost converter · SWT · DC microgrid

1 Introduction

Small wind turbines (SWTs) are defined as turbines which do not exceed 50 kW and are mainly found installed close to inhabited areas where large scale wind turbines cannot be erected. In countries where high wind speeds are common, installation of SWTs in remote sites (or for off-grid applications) are a viable option even though they have high installation and maintenance costs. Such sites normally consist of a small group of buildings located relatively close to each other, which makes the application of a microgrid system very attractive for their electricity supply. A microgrid system consists of a cluster of generation sources (such as photovoltaic, wind turbines and even fuel based generators) and electrical loads which can function both connected to the main electrical grid as well as independently from the supply network. The microgrid's control aims at maintaining a constant voltage under all operating conditions.

D. Zammit (✉) · C. Spiteri Staines · A. Micallef · M. Apap
Department of Industrial Electrical Power Conversion, University of Malta, Msida 2080, Malta
e-mail: daniel.zammit@um.edu.mt

L. Battisti and M. Ricci (eds.), *Wind Energy Exploitation in Urban Environment*, Green Energy and Technology, https://doi.org/10.1007/978-3-319-74944-0_14

There exist two types of microgrids, mainly; AC and DC. An attractive aspect of microgrids is the ability to operate both in grid-connected mode as well as in islanded mode. In grid-connected mode the microgrid is connected through a coupling point to the electrical grid, while in islanded mode the microgrid is operated in an autonomous way disconnected from the electrical grid. In this paper, the connection of a SWT to a grid-connected DC microgrid shall be discussed. DC microgrids offer a number of advantages, such as, lower conversion losses (fewer DC to AC or vice versa conversions), no synchronisation, phase or frequency issues (as present in AC microgrids), and also independence from voltage sags, dips, and other power quality issues occurring on the electrical AC grid side.

The most common configuration for SWT grid-connected systems of powers less than 20 kW normally consist of a high pole pair permanent magnet synchronous generator (PMSG) feeding a three phase rectifier followed by a boost converter and a grid-connected single phase AC inverter. When incorporating a SWT to a DC microgrid, the grid-connected inverter can be omitted, thus decreasing the power conversion system cost by more than half, while at the same time increasing the efficiency of the overall system. The most critical and complex part of the standard AC grid-connected system is the control of the boost converter which is used to regulate the voltage amplitude at the terminals of the PMSG allowing for maximum power transfer. Many wind turbine (WT) manufacturers provide a wind turbine power curve to be used for the MPPT control, which is normally inbuilt in the grid-connected inverter. However, unfortunately this characteristic can be inaccurate resulting in poor MPPT operation. In addition, the pre-programmed characteristic is fixed and hence does not cater for changes which occur over the years. Thus it is recommendable that the MPPT for wind based systems should either be adaptive [1] or independent of the WT characteristic [2, 3]. There are three types of MPPT systems in SWTs [2, 4]:

1. Power Curve Characteristic Control: which requires accurate knowledge of the turbine parameters, and requires power and rotational speed measurement.
2. Optimal Tip Speed Ratio Control (λ_{opt}): which requires knowledge of λ_{opt}, wind speed and rotational speed.
3. Perturbation and Observation (incremental MPPT) Control: which requires no knowledge of the turbine characteristics and needs power measurement (and in some cases speed measurement).

The first two methods have got a better dynamic response making them suitable for situations of rapid wind variations. On the other hand, the third method is parameter independent, simple to implement, and can be applied more to a wide range of different types of SWTs.

This paper proposes an incremental MPPT control methodology for direct connection of a 5 kW SWT to a DC microgrid system, which includes current control. An incremental MPPT system covered in [5] was based on the direct variation of the duty cycle of the Boost converter according to the change in the DC link power. In this proposed MPPT algorithm the variation is made in the current, according to the change in the DC link power. This current is used as the reference to a Proportional-

Integral (PI) controller which controls the Boost inductor current, providing more controllability and robustness to the system.

The paper is organized as follows. Section 2 gives an introduction on wind energy conversion. Section 3 shows the considered system model, and describes the electrical conversion modelling and the MPPT algorithm. Simulation results for varying wind conditions are presented in Sect. 4.

2 Basics of Wind Energy Conversion

The power available in a wind stream P_{air} is given by [6, 7]:

$$P_{air} = \frac{1}{2}\rho A V_w^3 \tag{1}$$

where, P_{air} is the power available from the wind (W), ρ is the air density (kg/m^3), A is the cross-sectional area of the stream (m^2) and V_w is the free wind velocity (m/s). This power cannot be completely extracted and the conversion efficiency is dependent on the power coefficient C_P. The power coefficient expresses the ratio between the extracted power P_w and the available power P_{air} as defined by (2).

$$C_P = \frac{P_w}{P_{air}} \tag{2}$$

C_P can have a maximum of 0.593, known as the Betz limit. It is a function of two variables: the blade pitch angle β and the tip speed ratio λ. λ is defines as:

$$\lambda = \frac{\omega_{rot} R}{V_w} \tag{3}$$

where, R is the radius of the rotor (m) and ω_{rot} is the rotational speed of the rotor (rad/s). Hence, the power extracted by the rotor of a wind turbine is given by:

$$P_w = \frac{1}{2}\rho A V_w^3 C_P(\beta, \lambda) \tag{4}$$

In the case of the SWT, the pitch angle is fixed, thus C_P is solely a function of the tip speed ratio λ. A typical Cp-λ curve for a horizontal axis three bladed wind turbine is shown in Fig. 1.

By controlling the rotational speed ω_{rot} of electrical generator, the tip speed ratio can be adjusted to maximize the C_P. However as shown in Fig. 2, the maximum power operation point depends also on the wind speed. For every different wind speed there is a different optimal tip speed ratio. By using variable speed operation of the electrical generator maximum power extraction can be achieved at different wind speeds.

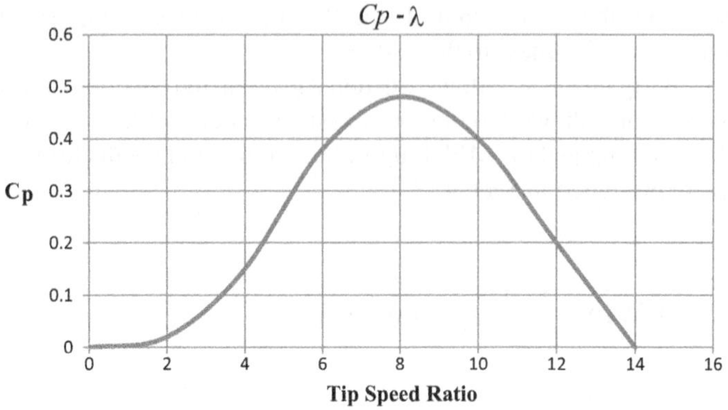

Fig. 1 Typical *Cp*-λ curve for three bladed WT

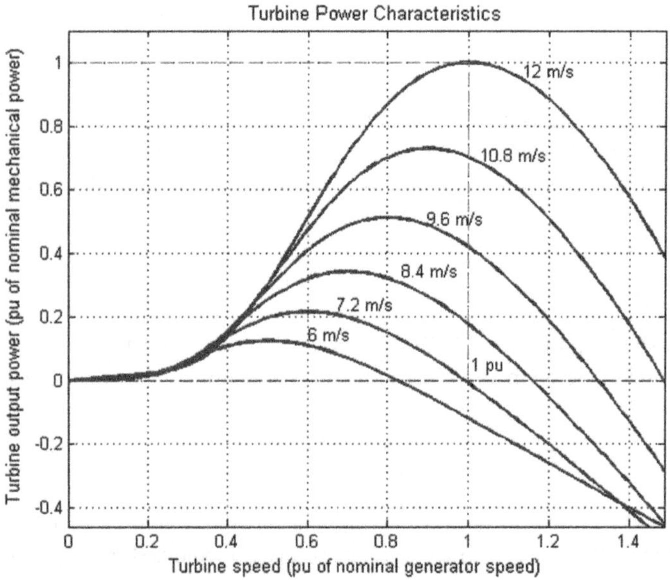

Fig. 2 Normalised extracted power curves for wind turbine

3 System Modelling

The considered DC microgrid setup is shown in Fig. 3. For the case being considered in this paper, the microgrid is assumed to be operating in grid-connected mode through the Grid Inverter. The SWT consists of a 5 kW PMSG feeding a 3-phase rectifier whose DC output voltage is controlled via a DC/DC converter. The latter converter has a boost topology and is used to interface the rectifier output to the DC

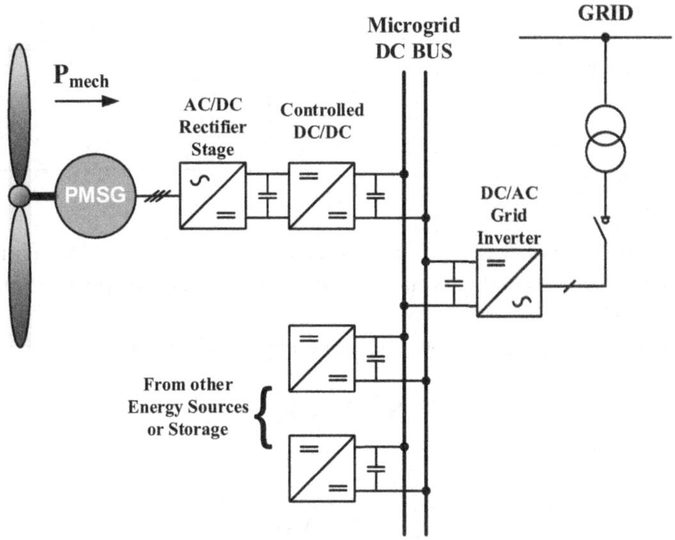

Fig. 3 Small wind turbine connected to a DC microgrid

microgrid. The boost converter was controlled to obtain maximum power from the SWT into the DC grid.

3.1 FMA Analysis of the PMSG

Figure 4 shows the most commonly used topology for the rectification of PMSGs used in small wind turbines. The three phase diode rectifier normally has a largely capacitive DC link which determines the voltage across the PMSG's terminals. Figure 5 shows the equivalent single phase circuit of the PMSG after applying fundamental mode approximation (FMA) to the terminal voltage phasor $\mathbf{V_{AN}}$:

$$\mathbf{V_{AN1}} = \mathbf{E} - \mathbf{Z_S I_S} \tag{5}$$

Fig. 4 PMSG feeding three phase rectifier with large DC link capacitance

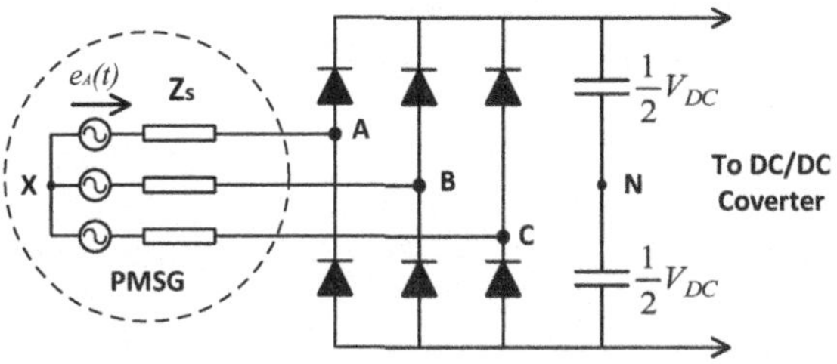

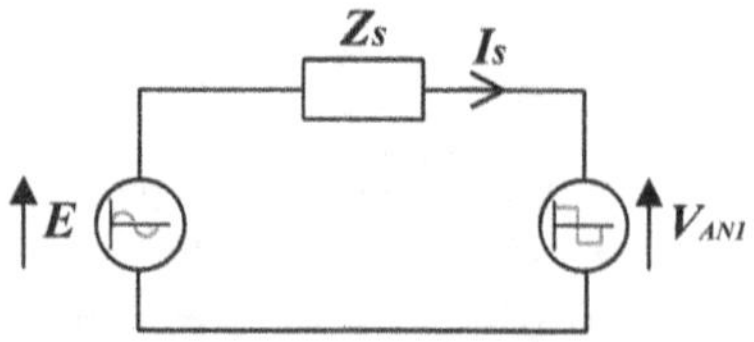

Fig. 5 FMA single phase equivalent circuit for PMSG with capacitive three phase rectifier

where, $\mathbf{V_{AN1}}$ is the fundamental voltage of the machine, $\mathbf{Z_S}$ is the PMSG impedance, $\mathbf{I_S}$ is the phase current, and $\mathbf{E}$ is the back-emf of the machine (proportional to the rotational speed).

The single phase apparent power generated by the PMSG is given by:

$$\mathbf{S} = \mathbf{V_{AN1}} \mathbf{I_S^*} \tag{6}$$

where,

$$\mathbf{I_S} = \frac{\mathbf{E} - \mathbf{V_{AN1}}}{\mathbf{Z_S}} \tag{7}$$

Substituting (7) in (6) it can be shown that the real power component is given by:

$$P = \frac{V_{AN1} E \cos(\theta_Z - \delta)}{Z_s} - \frac{V_{AN1}^2 \cos(\theta_Z)}{Z_s} \tag{8}$$

where, V_{AN1} is the RMS value of the fundamental component of the terminal voltage, E is the RMS value of the back-emf, and Z_s is the magnitude of the impedance. δ is the angle between the back-emf and the terminal voltage, and θ_Z is the angle of the PMSG impedance. In the case of low power PMSGs such as those used in small wind turbines, the PMSG impedance normally is predominantly resistive ($\theta_Z \approx 0$) and the real power can be approximated by:

$$P = \frac{V_{AN1}(E \cos(\delta) - V_{AN1})}{Z_s} \tag{9}$$

The total output power for the three phase machine is obtained by multiplying (9) by a factor of 3. Moreover from (9), one can note that the real power is dependent on the DC link voltage amplitude. The latter shall form the power control basis for the considered PMSG topology.

3.2 Boost Converter Topology and Control System Design

The boost converter shown in Fig. 6 is a switching converter that produces a higher average output voltage (DC microgrid voltage V_{dcgrid}) than the input voltage (DC link voltage V_{dc}).

For the boost converter, the voltage conversion ratio of the output voltage to the input voltage ($M(D)$), which is a function of the duty cycle (D) is given by:

$$M(D) = \frac{V_{dcgrid}}{V_{dc}} = \frac{1}{1-D} \tag{10}$$

The boost converter inductor (L) value can be found by:

$$L = \frac{V_{dc} D}{2\Delta i_L f_s} \tag{11}$$

where, Δi_L is the desired inductor current peak ripple, and f_s is the switching frequency of the boost converter.

The boost converter capacitor (C) value can be found by:

$$C = \frac{V_{dcgrid} D}{2\Delta v_o R f_s} \tag{12}$$

where, Δv_o is the desired output voltage peak ripple, and R is the load resistance. Since in this particular case R is not used in the simulation, it can be considered as the minimum load resistance to ensure Continuous Conduction Mode (CCM) operation of the converter.

To design the PI current controller, the transfer function of the duty cycle to the inductor current $G_{id}(s)$ should be used. This transfer function can be obtained by using the small signal equivalent circuit of the boost converter. The derivation of the boost converter small signal equivalent circuit and the transfer function are covered in [8]. In this case, the boost converter is being operated within a DC microgrid which is operated in grid-connected mode through the inverter. Hence, the transfer function of the duty cycle to the inductor current $G_{id}(s)$ can be reduced to:

$$G_{id}(s) = \frac{\hat{i}_L(s)}{\hat{d}} = \frac{V_o}{sL + R_L} \tag{13}$$

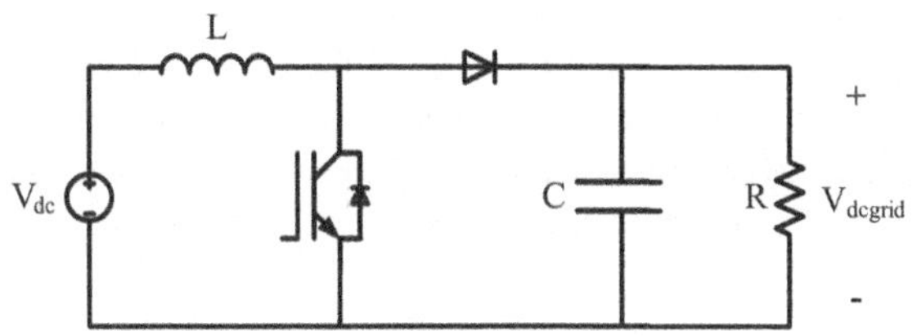

Fig. 6 Boost converter topology

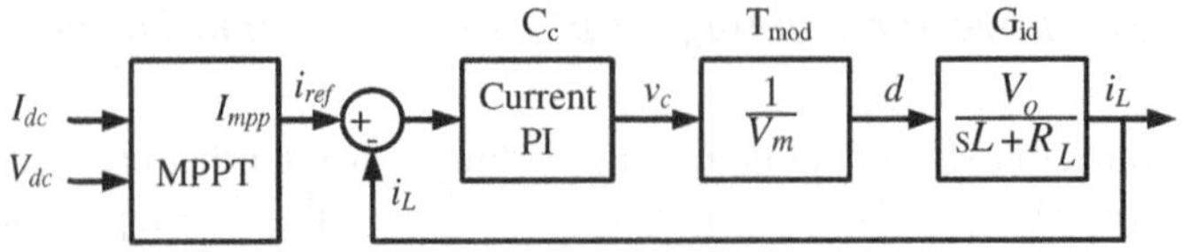

Fig. 7 Block diagram of the boost converter control system

where, $\hat{i}_L(s)$ and $\hat{d}$ are small ac variations around the quiescent values for the inductor current and duty cycle, respectively. L and R_L are the inductance of the boost inductor and the inductor resistance, respectively.

Figure 7 shows the block diagram of the control system including the MPPT block. The control system consists of a PI controller to control the inductor current, whose current reference value is obtained from the MPPT system. The MPPT algorithm uses the DC link current and voltage, I_{dc} and V_{dc}, respectively, as its inputs to provide the required MPPT current I_{mpp} as an output. The current I_{mpp} is used as the reference current, i_{ref}, to the current controller. In this paper, the DC link is being defined as the connection between the rectifier output and the boost converter input.

The current PI controller, $C_i(s)$ is of the form:

$$C_i(s) = K_p + \frac{K_I}{s} \tag{14}$$

where, K_P is the proportional gain term, and K_I is the integral gain term.

The plant transfer function $P_i(s)$ for the current PI controller is obtained by:

$$P_i(s) = T_{mod} \times G_{id}(s) \tag{15}$$

where, T_{mod} is the transfer function representing the pulse width modulation stage.

The pulse width modulation stage produces the duty cycle d that is proportional to the control voltage v_c. The pulse width modulator makes a comparison between the control voltage v_c and a sawtooth waveform with a peak to peak amplitude V_m. The value for V_m is selected by the designer. The frequency of the sawtooth waveform corresponds to the desired converter switching frequency f_s. The pulse width modulation stage can be modelled by the transfer function T_{mod} given by:

$$T_{mod} = \frac{1}{V_m} \tag{16}$$

The design of the boost converter as shown in Fig. 6 was done using (10), (11) and (12) to find the duty cycle D, the inductor L and the capacitor C, respectively [8]. The predefined set parameters and the calculated values are shown in Table 1. The inductor resistance R_L and the equivalent series resistance (ESR) of the capacitor R_c were taken from actual inductor and capacitor datasheets.

Table 1 Boost converter parameters

Set parameters	
Input voltage (DC link voltage) V_{dc}	50–339 V
Output voltage (DC grid voltage) V_{dcgrid}	400 V
Switching frequency f_s	10 kHz
Converter power P	5 kW
Inductor current peak to peak percentage ripple $2\Delta i_L$	0.01%
Output voltage peak percentage ripple Δv_o	0.5%
Inductor resistance R_L	0.002 Ω
ESR of capacitor R_c	0.03 Ω
Calculated values	
Duty cycle D	0.153–0.875
Inductor L	35 mH
Capacitor C	550 μF

The plant transfer function $P_i(s)$ was used to design the current PI controller gains in Matlab. The value for V_m was selected to be 440 V, a value larger than the maximum voltage in the system, in this case the DC microgrid voltage of 400 V. The values for the proportional gain term K_P and the integral term K_I for the current PI controller are 23.76 and 880, respectively. The bandwidth for the current closed loop which was designed is of 104 Hz.

3.3 *MPPT Algorithm for Small Wind Turbine*

The proposed algorithm only requires power output measurement for operation. This can be readily achieved from the DC link's current and voltage without requiring any speed or wind measurements. Maximum power operation for a wind turbine occurs when:

$$\frac{dP}{d\omega} = 0 \tag{17}$$

Moreover the back-emf of a PMSG is proportional to the rotational speed, ω, and hence it can be shown that:

$$\frac{dP}{dV_{dc}} = 0 \tag{18}$$

where, V_{dc} is the DC link voltage which is proportional to the amplitude of the PMSG's terminal voltage. Therefore, the power difference *dP* can be used to deter-

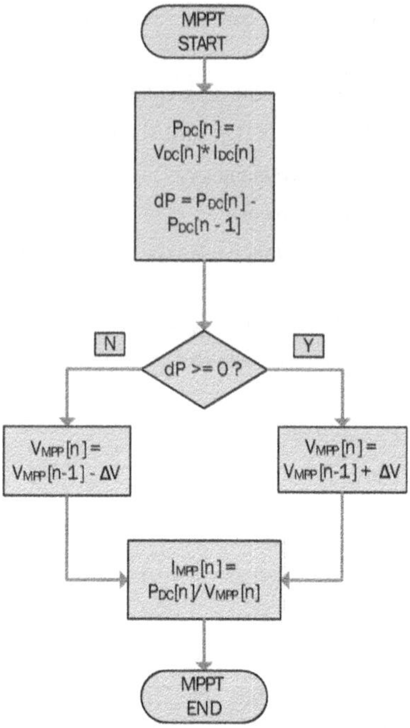

Fig. 8 Flowchart of MPPT algorithm for SWT boost converter

mine the need to increase or decrease the DC link voltage and consequently the DC link current to obtain maximum power at that particular wind speed.

Figure 8 shows the flowchart of the MPPT algorithm used in the simulations presented in this paper. This algorithm obtains maximum power by calculating the required current value I_{MPP} at that particular wind speed. I_{MPP} is then used as the reference for the current PI controller. The MPPT algorithm increases or decreases the internal reference MPPT voltage V_{MPP} in steps according to the increase or decrease in the DC link power P_{DC}. The selection of the step size constant ΔV is very important. A too small step size causes the MPPT algorithm to take longer to track the maximum power point. On the other hand, a large step size, although it accelerates the tracking, causes larger oscillations around the steady state operating point [2]. Therefore, a compromise for the step size value should be found between fast tracking and minimized oscillations.

4 Simulations

Simulation of the microgrid system was carried out for a 5 kW small wind turbine whose parameters are given in Table 2. A block diagram of the complete system is shown in Fig. 9. The wind turbine including the PMSG and the boost converter were modelled in Simulink, and are shown in Fig. 10. The simulation considered

Table 2 PMSG AND SWT parameters

PMSG parameters	
Rated power	5 kW
Stator resistance R_s	1.95 Ω
Stator inductance L_s	1.9 mH
Number of pole pairs	15
Rated speed	200 rpm
System inertia	1 kg/m^2
Viscous damping	0.024 Nms
SWT parameters	
Cut-in wind speed	2.5 m/s
Rated wind speed	12 m/s
Rotor diameter	5.5 m

the operation of the SWT with the microgrid operating at a fixed DC voltage V_{dcgrid} of 400 V under grid-connected conditions and with varying wind speed conditions. A typical wind speed profile from 6 to 12 m/s was provided to the wind turbine model with steps in wind speeds, shown in Fig. 11. Although a step change is not representative of actual wind speed changes, it provides the worst case for the tracking algorithm. The voltage step size constant ΔV was selected to be 0.1. The PWM frequency was set to 10 kHz.

The simulation results for the DC link output power, the DC link voltage, the DC link current, and the PSMG rotor speed are shown in Figs. 12, 13, 14 and 15. These results show the attainment of maximum power transfer as the wind speed varies using the proposed MPPT algorithm. Figure 12 shows the DC link output power. At wind speeds of 6 m/s, 8 m/s and 10 m/s, the output power achieved was about 700 W,

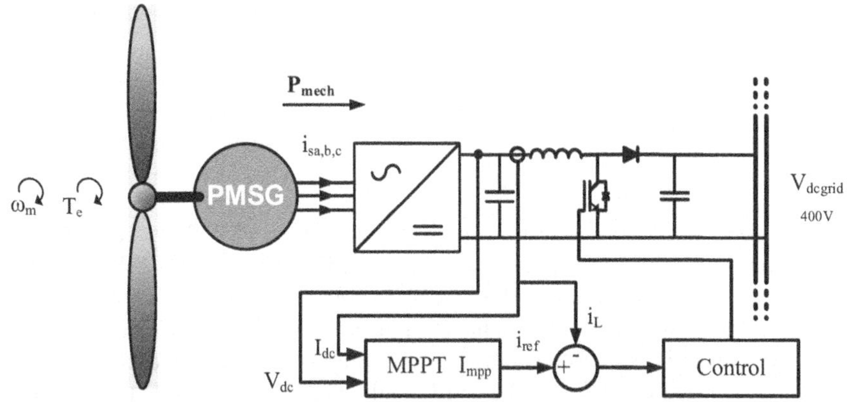

Fig. 9 Block diagram of the wind turbine including PMSG, boost converter and control system

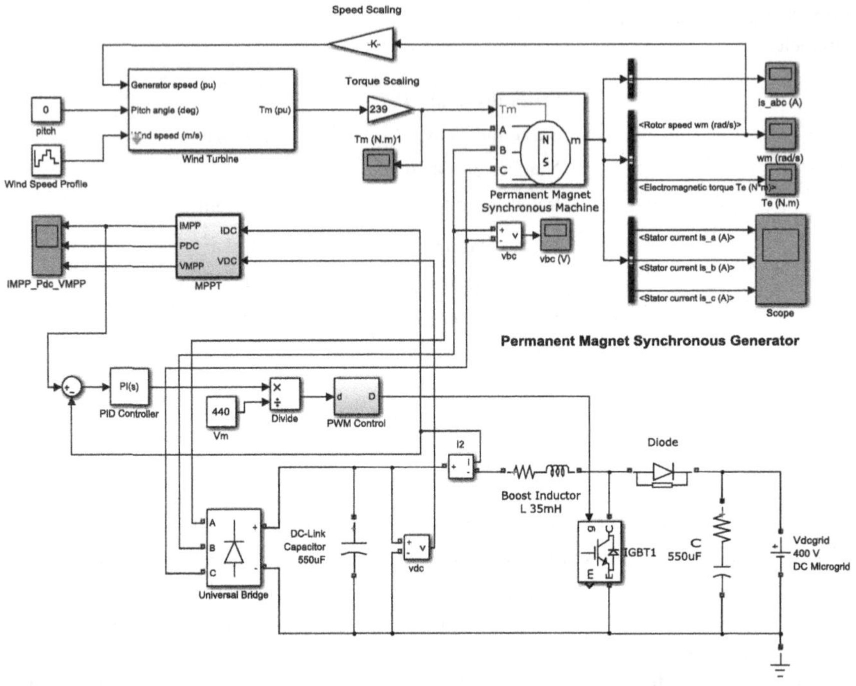

Fig. 10 Wind turbine including PMSG and boost converter simulink model

Fig. 11 Wind speed profile

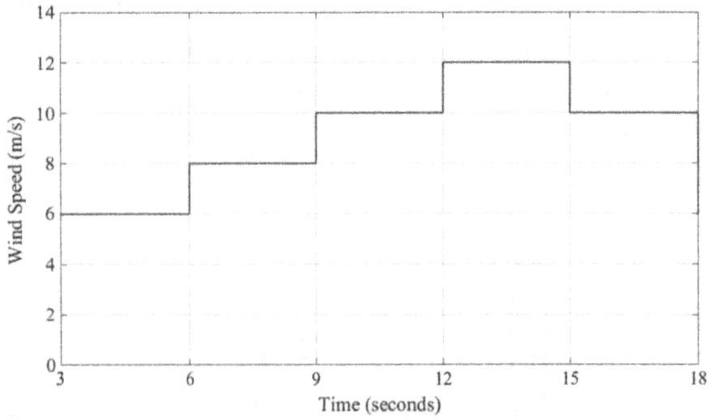

1.84 kW and 3.3 kW, respectively. The rated power of the PMSG was reached with a wind speed of 12 m/s, as expected, delivering a DC link power of 5 kW. Figure 13 shows how the DC link voltage before the boost converter changes with wind speed, while Fig. 14. shows the changes in the DC link current. As explained in Sect. 3, the current reference is increased/decreased according to the increase/decrease of the DC link power. Figure 15 shows the corresponding PMSG rotor speed for the above conditions. Figures 16 and 17 show the distortion in the stator phase current and the line terminal voltage of the PMSG, respectively.

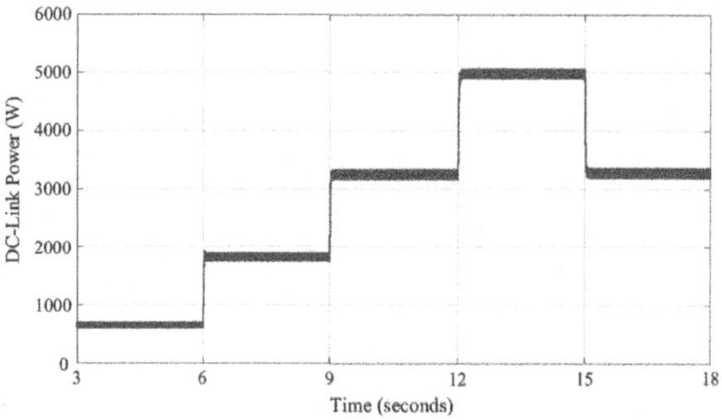

Fig. 12 DC link output power

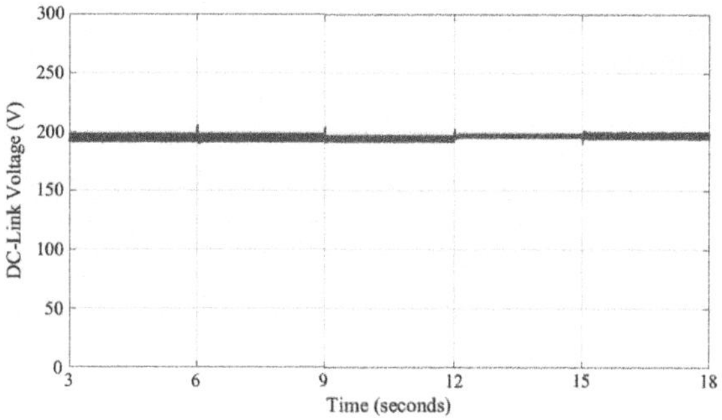

Fig. 13 DC link voltage

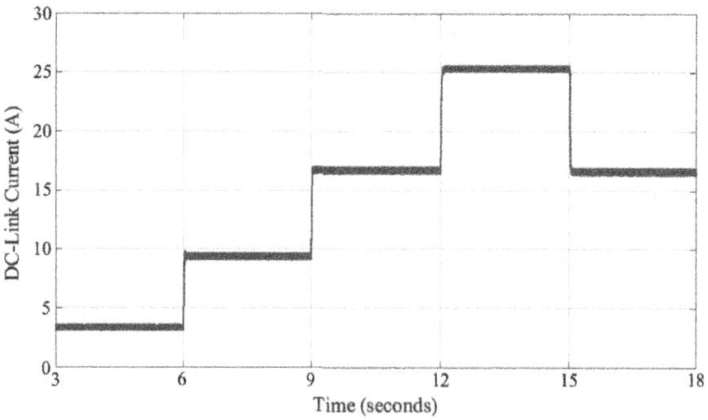

Fig. 14 DC link current

5 Conclusion

The paper presented an MPPT algorithm to obtain optimal power operation of a small wind turbine connected to a grid-connected DC microgrid. The system consisting of a PMSG driven by a SWT was interfaced to the grid-connected DC microgrid through a rectification stage and a boost converter. The proposed MPPT algorithm is based on the power-speed characteristic of the wind turbine system, but is independent of the system parameters. The proposed MPPT algorithm controls the boost inductor current according to variations in the DC link power, to obtain maximum power

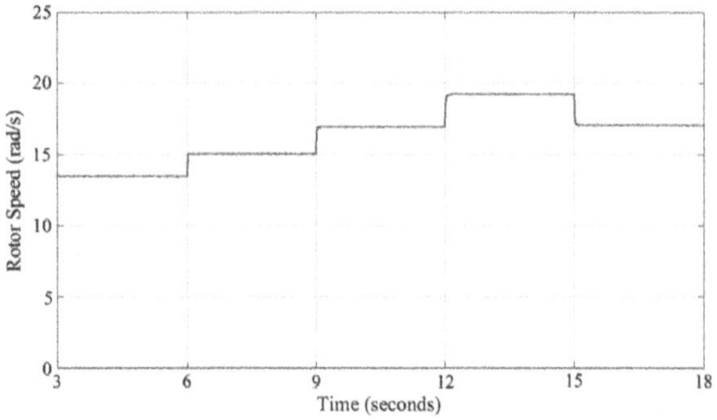

Fig. 15 PMSG rotor speed

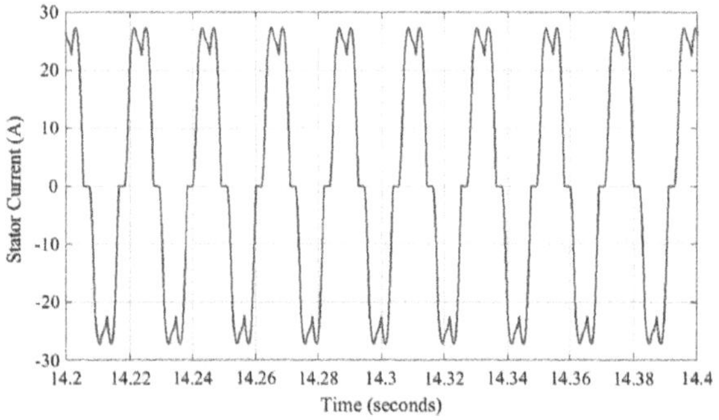

Fig. 16 PMSG stator current I_a

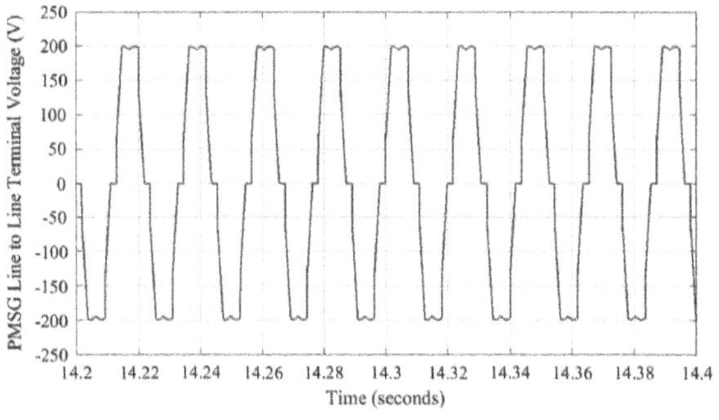

Fig. 17 PMSG line to line terminal voltage

output. This algorithm was successfully implemented in the simulation model, and simulations for varying wind speeds from 6 to 12 m/s were carried out. The boost converter allowed for successful transfer of energy from a DC link voltage of around 200 V to a fixed microgrid voltage of 400 V, with a varying DC link current of around 3.5 to 25 A.

The simulation results showed that the optimal power operating point was achieved for all wind speeds. This tracking algorithm provided maximum power output into the DC microgrid under all conditions.

References

1. Kortabarria, I., Andreu, J., Martínez de Alegría, I., Jimenez, J., Ignacio Garate, J., Robles, E.: A novel adaptative maximum power point tracking algorithm for small wind turbines. Renew. Energy **63**, 785–796 (2014)
2. Kot, R., Rolak, M., Malinowski, M.: Comparison of maximum peak power tracking algorithms for a small wind turbine. J. Math. Comput. Simul. (Elsevier) **91**, 29–40 (2013)
3. Aubrée, R., Auger, F., Dai, P.: A new low-cost sensorless MPPT algorithm for small wind turbines. In: 2012 First International Conference on Renewable Energies and Vehicular Technology, Hammamet, pp. 305–331 (2012)
4. Abdullaha, M.A., Yatima, A.H.M., Tana, C.W., Saidurb, R.: A review of maximum power point tracking algorithms for wind energy systems. Renew. Sustain. Energy Rev. **16**(5), 3220–3227 (2012)
5. Zammit, D., Spiteri Staines, C., Micallef, A., Apap, M.: Optimal Power Control for a PMSG Small Wind Turbine in a Grid-Connected DC Microgrid. (unpublished)
6. Burton, T., Sharpe, D., Jenkins, N., Bossanyi, E.A.: Wind Energy Handbook, 2nd edn. Wiley (2001)
7. Manwell, J.F., McGowan, J.G., Rogers, A.L.: Wind Energy Explained, 1st edn. Wiley (2002)
8. Zammit, D., Spiteri Staines, C., Apap, M., Micallef, A.: Overview of Buck and Boost converters modelling and control for stand-alone DC microgrid operation. In: Offshore Energy and Storage Symposium (OSES 2016), Valletta, Malta, July 2016

Landscape Compatibility of Wind Energy: Suggestions from Italian and Foreign Guidelines

R. Laviscio

Abstract The need of a widespread landscape quality (European Landscape Convention, 2000 [7]) has led, in Italy as abroad, to the establishment of laws, procedures and tools for a preliminary assessment of territorial transformations in relation to their landscape compatibility. The awareness of the ineffectiveness of a purely prescriptive rule has led to a wide production, since some decades, of "guidelines" as a tool to accompany design activity with qualitative suggestions, stemming from a careful reading of the landscape in which the project is placed. The paper analyses guidelines for wind farms, referring, in particular, to micro wind turbines and the urban environment. It wants to highlight their structure, contents and methods of communications. Through the comparison of national and international experiences the paper reflects, first, on suggested methodologies for landscape knowledge: considered issues, suggested analyses, study scales, data and knowledge sources; then, on suggested design criteria: considered aspects (e.g. materials, colours, localization), sources from which they came from (e.g. best practice, theoretical studies), representation modes (e.g. drawings, sketches, photos). Nevertheless, the paper points out possible shortcomings and reflects on the policies and tools necessary for guidelines' dissemination in order to guarantee a more effectiveness in practice. Research was conducted by exploring literature on the subject and web resources. In particular, the Italian context has been explored consulting regional and national guidelines provided by institutional bodies. The main international guidelines on wind energy has been explored too, allowing to compare the Italian situation with a broad context.

Keywords Landscape compatibility · Landscape preservation · Landscape knowledge · Guidelines · Wind turbines · Urban environment · Italy

R. Laviscio (✉)
Department of Architecture Built Environment and Construction Engineering, Politecnico di Milano, Via Bonardi 9, 20100 Milan, Italy
e-mail: raffaella.laviscio@polimi.it

L. Battisti and M. Ricci (eds.), *Wind Energy Exploitation in Urban Environment*, Green Energy and Technology, https://doi.org/10.1007/978-3-319-74944-0_15

1 Introduction. Landscape Compatibility: A Key Question in Territorial Transformations

The increasing awareness of the problems associated with the use of fossil fuels has, for the last twenty years, oriented Italy towards renewable sources, especially wind. This environmental requirement, arising from Italy's signing of the Kyoto Protocol, must compare, in our country, with the design of new energy landscapes in a context characterized by extraordinary landscape values. Hence the need to regulate the installation of wind turbines in relation to the impacts they may have on the landscape.

The practice shows, in fact, how large-scale wind farms have indiscriminately spread, regardless of the reference contexts and without a design quality capable of creating new landscapes. This is a problem that does not yet involve micro-wind turbines, uncommon in Italy, but that seems urgent to address to avoid effects already experienced in the telecommunications sector, where proliferation and poor quality projects have determined quality shortage of many places.

Moreover, the poor development of micro wind power in Italy, as in other countries, is due to the low ability to manage their landscape impacts, particularly in the urban environment, in areas of historical and environmental value [2].

The need of a widespread landscape quality [7] has led, in Italy as abroad, to the establishment of laws, procedures and tools for a preliminary assessment of territorial transformations in relation to their landscape compatibility [8].

The awareness of the ineffectiveness of a purely prescriptive rule has led to a wide production, since some decades, of "guidelines" as a tool to accompany design activity with qualitative suggestions, resulting from a careful reading of the landscape in which the project is placed.

The landscape project is, in fact, the only opportunity to highlight the specificities of the places object of transformations, whether it is a conservative approach, adapting as much as possible to the contextual features, through the research of the minimum possible incidence of the new project; whether it is an innovative approach, reinterpreting carefully contextual relations and characters and proposing new significant references able, however, to respect the consolidated values.

The reference landscape and the type of plant suggest the best approach; in both cases, however, a careful design is required that considers multiple factors from the choice of location, to the positioning of the different components, to the choices of materials and colours, or the arrangement of the surroundings, etc. More the project is going to be innovative, more it will have to understand the changes introduced in the pre-existing palimpsest and take care of sharing the new values introduced.

After all, the acknowledgment of the landscape value of an area is based on what is perceived and attributed by population according to their culture in each period [7].

Thus, landscape scenarios can be modified over time by community's perception.

An effective landscape approach implies a dynamic management, able to fit to the social context and not limited to passive limitations to design [15].

Guidelines could be a useful tool in this direction when they are the result of the integration of scientific knowledge and practices. Their aim is, first of all, to develop a landscape sensibility.

The paper analyses guidelines for wind farms, referring, in particular, to micro wind turbines and the urban environment.

Through the comparison of national and international experiences the paper reflects on suggested methodologies for landscape knowledge and design criteria: considered aspects (e.g. materials, colours, localization), sources from which they came from (e.g. best practice, theoretical studies), representation modes (e.g. drawings, sketches, photos).

Due to their constant updating they allow to reflect on current trends in raising awareness on landscape design of wind turbines, social acceptance of the theme, policies and tools implemented at different scales.

Nevertheless, the paper points out possible shortcomings and reflects on the policies and tools necessary for guidelines' dissemination in order to guarantee a more effectiveness in practice.

Research was conducted by exploring literature on the subject and web resources. In particular, the Italian context has been explored consulting regional and national guidelines provided by institutional bodies. The main international guidelines on wind energy has been explored too, allowing to compare the Italian situation with a broad context.

2 Guidelines as a Tool for Landscape Design of Wind Turbines

2.1 General Characters of the Guidelines

The theme of landscape compatibility of wind energy is, therefore, essential to all project scales.

In Italy, at a national level, the legislator first dealt with the issue through the Ministerial Decree of 10 September 2010 "Guidelines for the authorization of plants powered by renewable sources".

Referring primarily to wind farms, the guidelines set out the criteria by which Regions should identify measures for an appropriate landscape integration of the plants, to delimit areas that are not suitable for specific plant types and to define compensatory measures.

In particular, they suggest the reuse of areas already affected by human activities such as industrial sites, quarries, landfills, brownfield.

They invite to a design related to the specificities of the project area. Particularly with respect to agricultural areas, the integration of the plant is important in the context of local food and rural traditions, both in terms of its realization and its operation.

They are looking for research and experimentation of innovative design solutions, capable of ensuring the best possible integration of plants in historical, natural and landscape contexts.

They invite to citizen involvement in the planning and design stages and to a conscious information before the realization of the plant.

Still at a national level, the Heritage and Landscape Code (Law no. 42/2004) has introduced the "Landscape Report" (DPCM 12 December 2005) as a necessary tool for the authorization of transformation projects in listed areas. In consequence, the Ministry of Culture has proposed specific guidelines for wind energy [11] that suggest methodological criteria to guarantee landscape compatibility of wind projects considering the specific characters (morphological, formal, historical, perceptive, etc.) of each site. They represent a reference both in the formulation of regional-scale guidelines and in design practice; however, they refer to large plants and do not take into account the micro-wind technology (at that time not emerging in Italy).

The Regions are the main promoters in recent years of directives and guidelines for renewable energies both as a tool for design support and decision making.

These are heterogeneous documents [5, 29] that in most cases only secondly consider the landscape aspects and have the aim, instead, of streamlining the process, developing an implementation tool according to the Regional Energy Plan (where approved), defining the tools to ensure that the authorized plants are actually implemented, clarifying to the stakeholders the criteria and parameters on which assess the environmental compatibility in accordance with the current rules [1, 20, 22].

Regional guidelines are generally monothematic and therefore devoted solely to wind energy; except for Puglia and Lombardy who also devise guidelines for other types of renewable energy. Except for Puglia and Lombardy, the guidelines do not relate to regional landscape planning tools; they are more frequently linked to the Regional Environmental Energy Plan.

There is also a case of guidelines developed at a provincial level (Provincia di Foggia) in a close relationship to the landscape of competence.

The production of the guidelines is quite diffused also abroad: among others, France [16], Scotland [25, 26], Denmark [9], UK [27] and Australia [3] have done a wide work on wind farm providing methodological criteria both to read the landscape and to adopt design criteria related to it.

Some guidelines are specifically devoted to micro wind turbines and the urban context [4, 6, 14, 17, 18, 24, 25].

Foreign guidelines are produced by government agencies, but also by private associations or producers involved in sustainable development; they are dedicated to the theme of the wind in general, but also specifically to the micro-wind power and refer to the entire national territory.

In some cases, landscape is considered within the environmental aspects; in others, they focus directly on the landscape.

Some explicitly state the need to consult a landscape architect; others take into account aspects of population involvement in the design phase. They are generally addressed to technicians, administrators, but also to final consumers.

More frequently than the Italian ones, foreign guidelines are subject to updating. They advocate the use of bad and best practices and guide the reading with the development of an iconography rather substantial.

Brief guidelines devoted specifically to the landscape design of mini and micro turbines derive from some scientific research [13, 28].

2.2 *Landscape Analyses*

The heterogeneity of the guidelines produced in the different countries in relation to wind power is a sign of the many aspects that need to be considered in designing a wind plant that takes into account the landscape. There is widespread awareness that any intervention that modifies the configuration of a place, even of small size, such as mini and micro-wind turbines, must be considered a landscape project or should start from a careful consideration of the relationship between the plant and the reference context [4, 10, 12, 19, 21, 23].

Siting, design, materials and colours of machines, arrangement of surrounding areas will have to be the result of an organic project that knows how to enhance the pre-existence and bring added value. It is necessary to have an adequate knowledge of the place where it operates.

The design of single machines or small groups of machines generally linked to single utilities cannot avoid a careful monitoring of the context conditions and in particular of the fragmentation of the agricultural area, the presence of watercourses, vegetation, more or less notable and historical buildings to which the intervention is linked. If, in fact, the most important landscape impact is related to visual perception due to the need to install the turbines at a certain height and to the movement of the blades, however, the alterations of the systemic and symbolic relations among the different elements of the local and supra-local landscape cannot be neglected, alterations that generally increase in proportion to the number of installed turbines.

Schematically the impact on the landscape of wind turbines, even of a small dimension, can be recognized in:

- visual intrusion: appearance of new forms in "vertical" competition with other consolidated elements or isolated in landscapes from predominantly horizontal development;
- modifications of existing alignments and patterns;
- alteration of the social perception of the place.

If most studies insist on the centrality of visual analysis and requires the definition of the "visual influence zone" that points out where the turbines can be seen, other studies [10, 12, 14, 19, 23] emphasize the need to consider possible interferences with:

- historic or traditional centers of recognized importance;
- panoramic and landscape paths, belvedere and sensitive views;

- areas of high natural value, listed landscapes;
- landscapes characterized by a high degree of integrity, recognizability and notoriety;
- valuable agricultural areas.

Concerning specifically urban environment, there has been limited research into urban small scale wind energy [4].

Many highlight the economic inefficiency of this energy resource in urban environment due to the many interferences and turbulences caused by the concentration of different buildings and structures of different heights. Key characteristic of this landscape is a predominance of urban complexity, industrial buildings, infrastructure or a combination of these [10].

Most guideline underline the necessity to survey possible obstacles nearby the turbines.

In fact, where a site has obstacles nearby, a turbine needs to be sited higher than those obstacles, with more impact on the amenity of a place [4].

Anyway, also in urban environment visual analyses are not enough to guarantee an effective landscape project.

Before proceeding with a wind installation, it is necessary to proceed with an accurate detailed survey not only to detect the technical capacity of the wind, but also the layout and height of the buildings, the relationship between closed spaces and open spaces, if the site is surrounded by ground, obstacles such as buildings and trees, and the heights of these obstacles, the presence of buildings of interest, the presence of public areas, or places of collective memory.

Not many studies specify how these analyses can carried out, sources and representation moods.

2.3 Design Criteria

By referring explicitly to mini and micro wind turbines (both to small isolated turbines and to small groups of small-size machines tied to single utilities), Italian and foreign guidelines dictate some landscape criteria to minimizing impacts, if inevitable.

An accurate landscape design of mini and micro wind turbines in urban environment depends on the following factors.

Site

Site assessment is one of the key factors in the successful installation of small and micro wind turbines. It may play an effective role in reducing visual impacts and improve the general perception related to the technology, making it more acceptable to the public [14].

Thus, a proper site assessment is a difficult process that includes, as well as the wind resource assessment, the evaluation of the landscape characters of the site [18].

The siting of the turbines have to prefer locations where possible amenity impacts can be mitigated; locations as habitats of sensitive species, visually sensitive land-

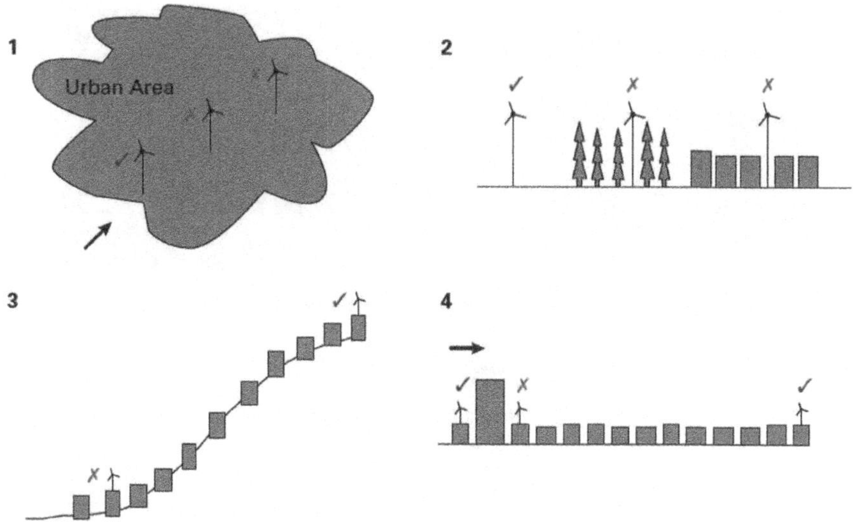

Fig. 1 Deciding where a site is suitable: in urban context turbines should be sited near the edge of built up area [4]

scapes, townscapes and buildings of architectural/historical importance are considered, in most cases, inappropriate [14, 19, 21].

Instead, some of the best urban wind resources are to be found along the edge of the built up area and atop tall buildings [4, 17] (Fig. 1).

In peri-urban areas the presence of other infrastructures such as power lines and towers, agricultural buildings and roads may create a context that visually absorbs a wind energy development. Visual complexity of industrial structures, for example, can help to assimilate a wind energy development where it becomes just one more element in the landscape.

A useful support can be tall elements as tertiary buildings, chimneys, water towers, gas meters, silos, sheds, electricity plants, urban furnishing items [19].

Anyway it is necessary to avoid the creation of visual confusion and spatial dominance where landscapes are already cluttered, but take advantage of a moderate amount of visual absorption that might be provided by existing structures and/or infrastructure [10].

The essential key is order and simplicity. Wind energy developments can thematically complement the contemporary technology expression of these landscapes and visually relate to their functional nature [10].

In case of placement on ground-based supports, it is opportune to favour, as far as possible, infrastructural placements: streets, railway lines, technological corridors [19] (Figs. 2 and 3).

Fig. 2 Sensitive siting: roads and galleries are a suitable site for the installation of micro wind turbines well integrated in the structures [19]

Fig. 3 Sensitive siting: micro wind siting seems more appropriate on buildings and structures that already have a technological character or at their open spaces such as parking areas at commercial or industrial centres [23]

Land form and pattern

The study of topography is essential to evaluate both the strength of the wind and the existence of any obstacles and the visibility of the plant in relation to the accessibility and the presence of natural and built visual barriers (Fig. 4).

The siting of single turbines or groups of small turbines must be adapted to the morphological characteristics of the site and of the visibility they produce: therefore, in a flat condition the prevailing view from the bottom should be considered, in hill or mountain conditions the view from high points [19].

Ideally turbines are located high up to take advantage of the prevailing winds.

Anyway it is opportune to take advantage of the possibility of a relationship between a wind energy development and a settlement, following its pattern: thus, for example, wind turbines can be arranged regularly in a line parallel to a straight field boundary or a road [4, 6, 19, 26].

Fig. 4 Land form: wind turbines reflect the existing pattern of built development by following settlement location, elevation and land form [6]

Fig. 5 Patterns: wind turbines follow the existing patterns of the landscape [26]

Also the height of towers should relate to the scale of landscape and the height of existing vertical elements such as light columns, telegraph poles, trees, buildings and other structures [26] (Fig. 5).

Built up

The placement of micro wind turbines in relation to existing buildings requires special attention. It has to guarantee respect for its dimensional, morphological and stylistic characteristics, taking into account, for example, the full-empty relationship, the presence of decorative or symbolic elements, any existing symmetry [14, 19] (Fig. 6).

Roof mounted turbine have not to detract from the architectural integrity of the building, especially if the building has conservation value. Anyway, the guidelines do not strongly recommend (or forbid) the installation on buildings of historical and/or architectural value.

Siting on a roof should consider the relationship to other existing roof structures such as chimneys, aerials, masts equipment housing, or where they can follow the form and positioning of traditional elements on roofs like chimneys.

Appropriate placements are behind cornices and parapets of flat roofs.

Anyway, instead of positioning on the highest point of the roof, to reduce their visual impact, guide lines suggest to locate wind turbines at a lower position provided they have a sufficient clearance from the base of roof.

Fig. 6 Placement on flat roof of a residential building in accordance with the character of the building [23]

Fig. 7 Positioning of turbines with respect to the symmetry character of the building [19]

In urban context the better positioning is at the rear of the property to minimise visual impacts from street level but care needs to be taken when back elevations are also visible from frequently accessed public recreational spaces.

Favourable placement is on accessory structures (garage and similar) when not visible from the road.

Where more than one turbine is proposed the aim should be to achieve a balanced composition, by grouping them together or installing through the use of symmetry (Fig. 7).

In some circumstances it may be better to site a turbine on a tower in the land adjacent to a building, particularly where a roof mounted turbine will damage the building's architectural integrity.

Turbines' height should relate to the existing buildings and vertical elements on site and the surrounding landscape [26].

Fig. 8 Integration of wind technology into a 0-emission housing project [23, 28]

Also tower mounted turbines would not integrate easily on historic buildings and their open spaces that are, in most cases, as valuable as the built structure itself and contribute towards their conservation value [14].

Different is the case of newly constructed buildings and complexes where the installation of the micro wind turbines can be suitably and totally integrated [23, 28] (Fig. 8).

Materials and colours

The colour and finish of the wind turbines need also to be taken in consideration and should be appropriate to the surroundings to minimise visual impact and reflection of light.

The adoption of different colours for micro wind turbines, as a mitigation measure against visual impact, depends on the view point from which the turbine is seen. Anyway, within most environments, neutral pastel colour as light no reflective grey, is highly adaptable to most lighting and weather conditions [14].

Some guidelines recommend specific colours depending on the background: an off-white colour may be appropriate where the main background is the sky, while green/brown colour may be suitable if the main background is vegetation or building elements [26].

Bright or dark colours as well as black or white should be avoided as these tend to stand out more [14].

In all cases, it is recommended the use of turbines with colours similar to the support and of no reflective colours, and to avoid affixing commercial logos [19].

Design

Aesthetic considerations represent some of the most critical issues in relation to wind energy development [30].

Compositional balance and harmony, rhythm, aesthetic order and clarity are essential to a positive perception of wind energy developments.

Concerning spacing, guidelines generally recommend a uniform type in any given wind energy development, rather than a mixture. Regular spacing is more appropriate

Fig. 9 Design: several designers have applied to the issue of micro wind turbines for domestic use. The image show a model based on a modular structure, an open and flexible "add-on" system made of mini windmills, which works like a LEGO puzzle. It may be located between buildings or over them being part of the whole construction (http://www.industrial-design-germany.com/innovations/wind-energy-system-windflock.html)

for wind energy developments in landscapes of clear and orderly land cover pattern; irregular spacing is more appropriate in landscapes of varied land cover pattern; graduated spacing of turbines is acceptable for wind energy developments where accentuation of a landscape feature is sought.

The same considerations apply to turbines' heights: a single height of the machines is preferable. This should be seriously considered when a new machine is added to an old plant. A wind energy development comprising two distinct turbine heights may be acceptable provided the resulting composition is carefully considered, so as to achieve an aesthetic effect.

If the choice of the machine may be of secondary importance in rural areas, the design becomes crucial in an urban environment where the system must necessarily deal with the existing [4].

Turbine types for urban application include horizontal and vertical axis wind turbines. Vertical axis turbines are often preferred for urban application for perceived aesthetic advantages, low noise levels and improved performance in turbulent flows [17].

Where possible, an innovative design should be adopted to create new positive landmarks (Fig. 9).

Design machines should be thematically related to different kind of landscapes and buildings when roof mounted, strengthening a certain identity (industrial contexts, infrastructures, modern buildings, open green spaces) (Fig. 10).

In most cases the use of a single type of machine is recommended within the same view .

Fig. 10 Design: the choice of the machine has to be related to the characters of the building of support [28]

Cumulative effects

The potential visual and amenity impacts are likely to be greater when micro wind turbines are located in the vicinity of each other [19]. This issue is significant for roof mounted micro wind turbines when located in densely developed urban areas and for tower mounted turbines too [14]. The location of more than one turbine on any particular site needs to consider the use of turbines of a similar form, design, colour and scale of those already existing.

Particular attention must also be paid to the placement of a new plant in relation to existing plants and significant landscape elements ensuring the maintenance of spatial relationships between [26].

Some conclusions and perspectives

The comparative analysis of Italian and foreign guidelines shows some shortcomings and some potentialities.

They are certainly a useful tool to ensure a wider landscape quality.

In fact, mostly foreign guidelines address designers but also final consumers. Expecting a greater development of micro wind technology in urban environment, this factor is very important: where, for small installations, the intervention of a landscape architect (in any case recommended) is not considered, the guidelines are a means of raising awareness of a landscape culture that anyone can understand.

Thus, the presence of a consistent iconographic apparatus and the use of bad and best practices is very useful, allowing a clear, quick and effective comprehension of the most appropriate way to operate depending on the context.

This mode that, at the moment, applies to the design criteria, should be also developed in relation to landscape analyses necessary to an appropriate wind energy installation.

In particular, referring to urban context there is some lack in guiding the reader to understanding its complexity, through what analyses and surveys, what instruments, what cartographic elaborations.

A further deficiency is to be found in the lack of integration of the guidelines in the territorial planning tools. It seems to be essential in prevision of a more extensive micro wind turbines development to avoid possible negative landscape impacts caused by uncontrolled dissemination of these plants.

Local governments, within their planning tools, may dictate priority localization and distribution rules, taking into account other existing or planned technology elements and specific local connotations; the wording of the criteria for the pooling and centralization of domestic plants should be addressed to them.

In this sense, the further development of the aspects dedicated to the involvement of the population in the wind energy design process (already developed abroad for wind farms) can have clear benefits from a landscape perspective, making community actions more effective than individual initiatives.

If the continuous updating of the guidelines (still to be pursued in Italy) and referring to current practices show current trends, the effectiveness of guidelines in practice is, nowadays, difficult to evaluate, lacking a monitoring system for interventions.

The introduction of a monetary and tax incentive mechanism for project compliance to the guidelines could be explored.

Guidelines are a tool of considerable interest that seems useful to continue to develop and deepen.

They help to define, in the design of technological infrastructures, a landscape culture.

The landscape project does not necessarily implies impress it a new sign; sometimes the minimum intervention criterion may be more appropriate.

A conservation approach to recognized values is not the sign of a passive attitude, but rather of conscious enhancement. Only from this, innovative landscape design can result.

References

1. APER Associazione Produttori Energia da Fonti Rinnovabili: Linee guida regionali per la realizzazione di impianti eolici e l'inserimento nel paesaggio (2011)
2. Attaianese, E., Duca, G., Caturano, U., Coppola, N., Scapicchio, S.: A hybrid micro generation system for a small scale building in urban area. Eur. Sci. J. **10**(23) (2014) ISSN: 1857-7881 (Print) e - ISSN 1857-7431
3. Australian Wind Energy Association, Australian Council of National Trusts: Wind farms and landscape values (2005)
4. Carbon Trust: Small-scale wind energy. Policy insights and practical guidance (2008)
5. Cinà, G.: Linee guida e progetto di paesaggio a scala regionale. Una sperimentazione in ordine sparso. In: Peano, A. (ed.) Fare paesaggio. Dalla pianificazione di area vasta all'operatività locale. Alinea Editrice, Firenze (2011)
6. Consultant Landscape Architects: Argyll and Bute Wind Energy Capacity Study. https://www.argyll-bute.gov.uk/planning-and-environment/landscape-wind-energy-capacity-study (2012)
7. Council of Europe: European Landscape Convention. http://www.coe.int/en/web/landscape/reference-texts (2000)

8. Cutaia, F.: Strategic Environmental Assessment: Integrating Landscape and Urban Planning. Springer (2016)
9. Danish Ministry of Foreign Affaire: Guidelines for the preparation and evaluation of investments in wind farms (2001)
10. Department of the Environment, Heritage and Local Government of Ireland: Planning guidelines. Wind energy development (2012)
11. Di Bene, A., Scazzosi, L. (eds.): Gli impianti eolici: suggerimenti per la progettazione e la valutazione paesaggistica. Gangemi Editore, Roma (2005)
12. ERM Environmental Resources Management Australia: Landscape and visual impact assessment (2008)
13. Laviscio, R.: Impianti di minieolico. In: Vallerini, L. (a cura di) Piano Progetto Paesaggio. Casa Editrice Pacini, Ospedaletto (PI) (2010)
14. MEPA Malta Environment and Planning Authority: Planning guidance for micro-wind turbines (2010)
15. Miccoli S.: Evaluation principles in Landscape Projects, AESTIMUM 61, Dicembre 2012, pp. 117–132
16. Ministère de l'Ecologie et du Développement Durable Agence de l'Environnement et de la Maîtrise de l'Energie: Guide de l'étude d'impact sur l'environnement des parcs éoliens (2004)
17. NSW Office of Environment and Heritage: NSW Small wind turbine consumer guide (2010)
18. Olsen, T., Preus, R.: Small Wind Site Assessment Guidelines (2015)
19. Regione Lombardia: Linee guida per le infrastrutture energetiche a rete. http://www.regione.lombardia.it/wps/portal/istituzionale/HP/DettaglioRedazionale/servizi-e-informazioni/Enti-e-Operatori/territorio/paesaggio/piano-paesaggistico-regionale/piano-paesaggistico-regionale (2010)
20. Regione Toscana, Piano di Indirizzo Territoriale con valenza di Piano Paesaggistico: Allegato 1b Norme comuni energie rinnovabili impianti eolici. http://www.regione.toscana.it/-/piano-di-indirizzo-territoriale-con-valenza-di-piano-paesaggistico (2009)
21. Regione Toscana: Linee guida per la valutazione ambientale degli impianti eolici (2012)
22. Regione Puglia: Linee guida per la realizzazione di impianti eolici nella Regione Puglia (2004)
23. Regione Puglia: Linee guida sulla progettazione e localizzazione di impianti di energia rinnovabile. http://www.paesaggiopuglia.it/pptr/tutti-gli-elaborati-del-pptr/4-lo-scenario-strategico.html (2015)
24. Scottish Executive Development Department: Planning for micro renewables. http://www.scotland.gov.uk (2006)
25. Scottish Natural Heritage: Siting and Design of Small Scale Wind Turbines of between 15 and 50 metres in height (2012)
26. Scottish Natural Heritage: Siting and designing wind farms in the landscape (2014)
27. Sustainable Development Commission: Wind power in UK (2005)
28. Urban wind turbines: Guidelines for small wind turbines in the built environment. http://www.urbanwind.net/pdf/SMALL_WIND_TURBINES_GUIDE_final.pdf (2007)
29. Vitale, Brovarone E.: Strumenti di indirizzo per gli impianti per la produzione di energia da fonti rinnovabili. In: Peano, A. (ed.) Fare paesaggio. Dalla pianificazione di area vasta all'operatività locale. Alinea Editrice, Firenze (2011)
30. Zanchini, E.: Paesaggi nel vento. Meltemi, Roma (2002)

The manufacturer's authorised representative in the EU is Springer Nature Customer Service Centre GmbH, Europaplatz 3, 69115 Heidelberg, Germany. If you have any concerns regarding our products, please contact ProductSafety@springernature.com

Printed and bound by CPI Group (UK) Ltd, Croydon, CR0 4YY
30/09/2026
02225601-0009